Software and Systems Architecture in Action

Software and Systems Architecture in Action

Raghvinder S. Sangwan

CRC Press
Taylor & Francis Group
Boca Raton   London   New York

CRC Press is an imprint of the
Taylor & Francis Group, an **Informa** business

AN AUERBACH BOOK

CRC Press
Taylor & Francis Group
6000 Broken Sound Parkway NW, Suite 300
Boca Raton, FL 33487-2742

© 2015 by Taylor & Francis Group, LLC
CRC Press is an imprint of Taylor & Francis Group, an Informa business

No claim to original U.S. Government works

Printed on acid-free paper
Version Date: 20140418

International Standard Book Number-13: 978-1-4398-4916-3 (Hardback)

Library of Congress Cataloging-in-Publication Data

Sangwan, Raghvinder S.
  Software and systems architecture in action / Raghvinder S. Sangwan.
    pages cm. -- (Auerbach series on applied software engineering)
  "A CRC title."
  Includes bibliographical references and index.
  ISBN 978-1-4398-4916-3
  1. Computer architecture. 2. Software architecture. 3. Computer systems--Design and construction. 4. Electronic data processing--Distributed processing. 5. Business enterprise--Computer networks--Design and construction. I. Title.

QA76.9.A73S29 2015
004.2'2--dc23   2014013462

Visit the Taylor & Francis Web site at
http://www.taylorandfrancis.com

and the CRC Press Web site at
http://www.crcpress.com

Contents

Preface

Modern-day projects require software and systems engineers to work together in realizing architectures of large and complex software-intensive systems. To date, the two have been using their own concepts, techniques, methods, and tools when it comes to requirements, design, testing, maintenance, and evolution of these architectures. This book looks at synergies between the disciplines of software and systems engineering and explores practices that can help software and systems engineers work together more effectively as a unified team.

The book illustrates an approach to architecture design that is driven from systemic quality attributes determined from both the business and the technical goals of the system rather than just its functional requirements. This ensures that the architecture of the final system clearly, and traceably, reflects the most important goals for the system. While superficially the most important goals of any system are its functions, in practicality it is the quality attribute requirements that have the greatest impact on a system's lifetime value because it is these requirements that determine how easily the system accepts future change and how well the system meets the reliability and security needs of its operators and owners. By making these requirements "first-class citizens," the architecture meets them first.

Furthermore, most quality attribute requirements are systemic properties: They are properties that the entire system must reflect rather than just one component or subsystem. They therefore cannot be easily built into an existing architecture. In essence, these properties must be designed into the architecture from the beginning. The architecture-centric design approach illustrated in this book addresses this directly by utilizing analytically derived patterns and tactics for quality attributes that inform the architect's design choices and help shape the architecture of a given system.

The book is organized into eight chapters. Chapter 1 focuses on the importance of architecture in modern-day systems. The amount and complexity of software in these systems are on the rise. Avionics software in modern aircraft has tens of millions of lines of code. In fighter aircraft, this software controls 80% of what a pilot does. It is not atypical for a premium-class automobile today to contain close to 100 million lines of code. The same is true for chemical and nuclear power plants. A large proportion of software in these systems introduce design and operational complexity, making them high-risk systems. This chapter highlights the role of architecture in managing this complexity and a need for an architecture-centric engineering approach to designing complex systems that can be used consistently by software and systems engineers working on such projects.

Chapter 2 looks at the influence of business goals or mission objectives on the architecture of a system. Business goals correspond to quality attributes the end system must exhibit. When using two different versions of a system, you may find them functionally equivalent but may develop a preference for one over the other because of its quick response time, ease of use, ease of modification, or high reliability. Such characteristics of a system are called quality attributes and are the predominant forces that shape the architecture of a system. Understanding business goals and their implied quality concerns is therefore critical.

A system operates within a given context or an environment, and understanding a system's operational aspects within its environment is also extremely important. Chapter 3 looks at the concept of operations or ConOps, a term used for the operational view of the system from the perspective of its users that gives a broad understanding of the capabilities a system must deliver to fulfill its mission objectives. An operational view helps clearly delineate where a system's boundary is, what elements in its external environment a system must interact with, and what those interactions are.

If one must wait for a system to be developed to determine if it will meet the quality expectations of its stakeholders, there is an inherent risk that it may not. Architectural restructuring to achieve the desired qualities at this stage may be extremely difficult and costly. It is much more desirable to predict the systemic properties of a system from its design so corrections can be made before the system is committed to development. Patterns are known solutions to recurring design problems and therefore have qualities that can help predict the systemic properties of the system that is built using them. In prescribing solutions to problems, patterns may use many design decisions. These design decisions are known as tactics. Patterns and tactics are topics described in detail in Chapter 4.

When designing a system, one often must consider several requirements that have a strong influence on its architecture. Many of these requirements frequently conflict with each other. For instance, while one requirement may need the system to be highly secure, another may need the system to have quick response time. Making a system secure may introduce authentication, authorization, and encryption mechanisms that introduce latency, thereby slowing the system. Chapter 5 explores an approach to creating an architecture that systematically addresses the architecturally significant requirements while also dealing with trade-off situations created by requirements that conflict with each other.

Architecture is an artifact that serves many diverse needs for many diverse stakeholders. For instance, project managers use it for organizing projects and distributing the work among development teams, teams use it as a blueprint for their development work and for understanding how their work depends on those of others, and maintainers use it to understand the impact of change as the system evolves over time. Effectively communicating the architecture to meet the diverse needs of a broad set of stakeholders is the topic of discussion for Chapter 6.

Once the architecture has been designed, the development teams must then undertake detailed design and implementation of the individual components that make up the final product. What is detailed design for one, however, is architecture to another. It turns out that effort must be invested by the development teams to develop the internal architecture of these components as they create the final blueprint for implementation. The interplay between architectural work and the detailed design is explored in Chapter 7.

Complexity is the topic of the final chapter, Chapter 8, which shows how following architecture-centric practices outlined in this text can lead to significant reduction in accidental complexity that is a by-product of development methodologies that lack focus on systemic properties of a system that have a strong influence on its architecture.

The fundamental objective of the book is to explore and illustrate practices that can be helpful in the development of architectures of large-scale systems in which software is a major component. It should be particularly useful to those currently involved in such projects and who are looking at more effective ways to engage the software and systems engineers on their teams. The book can be also used as a source for an undergraduate or graduate-level course in software and systems architecture as it exposes the students to concepts and techniques used for creating and managing architectures of software-intensive systems.

About the Author

Raghvinder (Raghu) Sangwan
is an associate professor of software engineering at Pennsylvania
State University. His work
involves design and development of software systems, their
architecture, and automatic and
semiautomatic approaches to
assess their design and code
quality. He has published several

papers in these areas. Prior to joining the Pennsylvania State
University, Raghu was a software architect at Siemens, where he
worked on large-scale systems in the domains of health care,
automation, transportation, and mining; many of these systems
were developed by teams geographically distributed around
the world. This experience resulted in his coauthoring the
Global Software Development Handbook and co-organizing the
first International Conference on Global Software Engineering
(ICGSE 2006), sponsored by the Institute for Electrical and
Electronics Engineers (IEEE). He also holds a visiting scientist
appointment at the Software Engineering Institute at Carnegie
Mellon University. He earned his PhD in computer and information sciences from Temple University and is a senior member of
IEEE and the Association for Computing Machinery (ACM).

Chapter 1

Architecture and Its Significance

1.1 Introduction

A system is a set of elements so connected or related that they perform a unique function that cannot be performed by the elements alone. The picture of a modern aircraft in Figure 1.1 illustrates this point. The aircraft has many constituent parts, but none of these parts by itself is capable of flight. Only when put together in a particular way do they enable an aircraft to fly.

We can examine other kinds of systems and make similar observations. A corporation is a social system that produces goods and services that individuals working for the corporation cannot produce individually. An individual is an animate system in which no single part of the body by itself can produce life. A clock is a mechanistic system; its individual parts together serve the purpose of showing time. An airline reservation system is an information system that as a whole manages flight reservations for passengers.

Figure 1.1.   Which part of the airplane flies?

An architecture of a system is fundamentally concerned with how a system is organized into its constituent elements and how these elements relate to each other to achieve a given purpose (Bass et al., 2003). Given this perspective, every system has an architecture, but a suitable architecture is one that enables a system to achieve the purpose for which it was created. For instance, we can put together the wings, fuselage, engines, and the landing gear of an airplane, but this would do no good until they are put together in a way that makes the aircraft fly. Systems are not limited to just the hardware, however, and can also include people, software, facilities, policies, and documents. All of these elements may be required to produce the desired system outcome; for instance, a successful flight of a commercial airliner from its place of origin to its destination is as much a result of the aircraft and its crew as it is of the ground crew and air traffic control.

> Architecture is fundamentally concerned with the organization of a system into its constituent elements and their interrelationships to achieve a given purpose.

1.2 Rising Complexity

The complexity of modern-day systems continues to rise. These systems need to be created quicker, and new features need to be introduced faster. There is an increasing need to

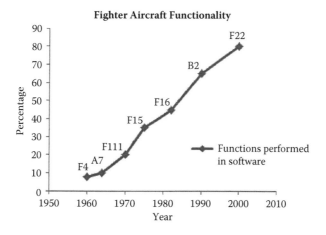

Figure 1.2   Proportion of software in modern-day fighter aircraft.

customize them for niche markets, and new requirements continue to surface so that systems must evolve to serve emerging
market needs.

Much of this required complexity has made it necessary to
incorporate a significant amount of software in many products. For instance, it is not atypical for a premium-class automobile today to contain close to 100 million lines of code.
The same is true for chemical and nuclear power plants. As
Figure 1.2 shows, avionics software in modern fighter aircraft
has steadily increased and today controls 80% of what a pilot
does (Reifer, 2001).

F-35 JOINT STRIKE FIGHTER

The Air Force Scientific Advisory Board has expressed
concern at the Air Force's ability to contend with the
growing utility and complexity of software in its aircraft.
At the initiation of the F-35 program in 2006, the esti-
mated size of software to be developed for the F-35 Joint
Strike Fighter aircraft was 6.8 million source lines of code.
More recent estimates have put this number at approximately 24 million. Meanwhile, the aircraft has surpassed

its delivery date by several years, and the Pentagon is expected to spend over $1 trillion on the development and sustainment of the F-35 program (Hagen and Sorenson, 2013).

The complexity is not necessarily related to the size of a system, but it is related to the interrelationships among the elements that make up such a system. Figure 1.3 shows a dependency graph of a system for viewing chemical structures. The elements of the system are represented as nodes, and their dependencies are shown as edges. As one can see, the number of elements and the dependencies among them is rather large. In its current form, it would be rather challenging not only to understand these dependencies but also to maintain and evolve this system over time. An architecture brings to

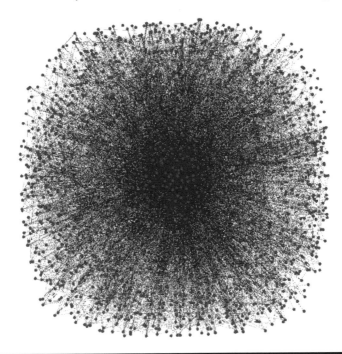

Figure 1.3   Dependencies among elements of a system for viewing chemical structures.

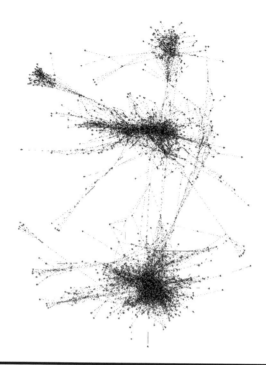

Figure 1.4   Elements reorganized based on how closely related they are.

bear organizing principles that can help manage this complexity, making a system not only intellectually graspable but also easier to maintain and advance over time. Figure 1.4 is a different rendition of the system shown in Figure 1.3 using some of these organizing principles.

As Figure 1.4 shows, the structure appears less overwhelming and more modular, with the elements that are closely related clustered together. Modularity, of course, is only one of the many organizing principles used when creating architectures of complex systems. Also, this exercise was an afterthought by which we took a disorganized system and restructured it to make it appear more modular. Architectures require much forethought and planning so that systems developed do not end up a complicated mess from the beginning.

Although some complexity is inherent in the problem domain for which a system is being designed and therefore, cannot be avoided, a system can be designed in a way that

SOFTWARE SYSTEMS ON AN AIRCRAFT

Known for its mechanical and software complexity, Airbus 380 is the largest civil aircraft built so far; it is put together from approximately 4 million parts from 1500 different companies and 1,000 onboard systems running several million lines of code (Burger et al., 2013). These include a high-lift system (HLS) for controlling all electrical and hydraulic functions within the wings; cabin intercommunication data system (CIDS) for controlling cabin-related functions; in-flight entertainment (IFE) system for passenger entertainment; airline network architecture (ALNA) for GSM and Internet connectivity; the air data and inertial reference system (ADIRS) for monitoring parameters such as speed, altitude, flight vectors, and so on; the flight management system (FMS) for optimizing travel routes to minimize flight time and fuel consumption; and the electrical flight control system (EFCS), which submits a pilot's control stick commands to the aircraft's control surfaces (e.g., ailerons, elevators, and rudders). The software complexity comes not only from the size and interactivity among these onboard systems but also from non-functional or quality attribute requirements especially those related to safety. All aircraft software is subject to FAA certification that uses the Radio Technical Commission for Aeronautics standard DO-178B, Software Considerations in Airborne Systems and Equipment Certification, which specifies different categories of software systems based on safety objectives they must satisfy (Avery, 2011). These are shown in Table 1.1.

makes it less complicated. Complexity that cannot be avoided is called incidental complexity, whereas complexity that can be avoided is termed accidental complexity (Brooks, 1995). The architecture of a system embodies design decisions that have an

Table 1.1   Aircraft Software Safety Levels and Categories

Safety Level	Category
A	*Catastrophic*: Failure may lead to a catastrophe such as a crash.
B	*Hazardous*: Failure may be hazardous to the safe operation of the aircraft, causing serious or fatal injuries.
C	*Major*: Failure is not hazardous but may lead to uncomfortable travel conditions for the crew and passengers.
D	*Minor*: Failure may inconvenience the crew and the passengers by causing flight delays.
E	*No Effect*: Failure is not related to the safe operation of the aircraft.

Source: Adapted from D. Avery, *IEEE Software* 28 (1), 11–13, 2011.

important part to play in effectively managing incidental complexity and avoiding accidental complexity.

> Architecture can help manage incidental complexity and avoid accidental complexity.

1.3 Constant Change

Complex systems are realized over a long period of time. This development time can see a lot of change in not only the technology used to create the system but also the requirements of the system itself. Figure 1.5 shows how mobile technology evolved from first-generation (1G) to fourth-generation (4G) systems in a span of 25 years; the 1G systems were analog and supported only voice communication, but the 4G systems are digital, ultrabroadband and support multimedia capabilities.

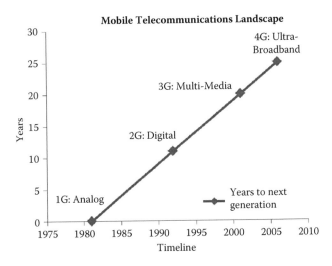

Figure 1.5   Changing mobile technology and its requirements.

In today's rapidly evolving world, change is the only constant; therefore, we must learn to embrace this change rather than try to control it. Architecture can play a key role in achieving this objective. A system can be designed with a flexible architecture that anticipates these changes. The value of creating such an architecture is the cost of not addressing the technical aspects that can help a system adapt to these changes when they happen. The Iridium fleet of communication satellites is an example of a system that was developed in the 1980s at a cost of about $4 billion and was launched in the late 1990s. Iridium phones cost as much as $3,000, with connection charges of up to $7 per minute, and failed in the short span of 9 months as they could not adapt in the face of less-expensive terrestrial cellular mobile phones that had swept the marketplace by then (De Neufville and Scholtes, 2011).

To mitigate the risk arising from constant change, most organizations develop systems incrementally over a number of iterations. Each increment produces a working fraction of a system that consists of a minimum marketable feature set and serves as a stable foundation for subsequent iterations. The architecture of a system plays a significant role in achieving this stability. It must anticipate the changes likely to happen in the foreseeable future and must accommodate these changes easily when they happen.

> The value of creating a flexible architecture is the cost of not addressing the technical aspects that can help a system adapt to changes when they happen.

1.4 Distributed Development

Increasingly, complex product development requires the use of expertise of many teams; these teams are often geographically distributed. This has given rise to the additional complexity of communicating, coordinating, and controlling work across geographically distributed development teams. As shown in Figure 1.6, the additional complexity results from the many dimensions of distance that come into play when project teams are far apart (Sangwan, Bass, et al., 2006). Physical distance limits face-to-face interactions, temporal distance limits overlap in working hours, cultural distance introduces communication difficulties caused by language and cultural barriers, and cognitive distance limits shared understanding. The most noticeable effect

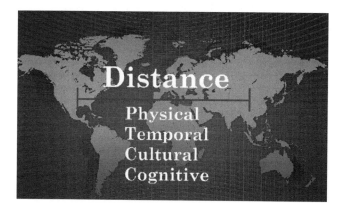

Figure 1.6   Distance in globally distributed software development.

of distance is a nearly total absence of informal or unplanned communication across the different sites, leading to difficulties in knowing who to contact about what, initiating contact, communicating effectively across sites, and establishing trust (Herbsleb and Grinter, 1999). This leads to conflicting assumptions or incorrect interpretation of communications across different sites, resulting in a delay in the resolution of work issues.

Constant change coupled with distributed development can create additional challenges. Change brings with it a lot of uncertainty and ambiguity. Although much of this uncertainty and ambiguity is addressed in colocated projects through informal communication among team members, geographically distributed teams do not have this luxury. Under such circumstances, architecture becomes the mechanism through which communication, coordination, and collaboration needs of distributed development projects are established. An architecture defines the relationship among the different elements of a system and captures their interdependencies. These interdependencies determine the volume, frequency, and type of communication, coordination, and collaboration needed among the participants in a globally distributed project. Architecture also helps create a shared level of understanding of the system context, its problem domain, and an overarching vision of the system to be designed. Establishing this shared context goes a long way in understanding dependencies across elements of the architecture and the allocation of their development to the distributed teams (Sangwan, Bass, et al., 2006). Lack of such a context leads to the creation of multiple realities promoting ignorance, confusion, and frustration, which subsequently undermine mutual trust and make inter-team communications less effective. This vicious cycle leads to dysfunctional teams, inefficiencies in the project, and ultimately poorly designed systems (Sangwan and Ros, 2008).

The architecture of a system therefore can exert a strong influence on the structure of a project. Figure 1.7 shows the structure of a system as a network of elements and their

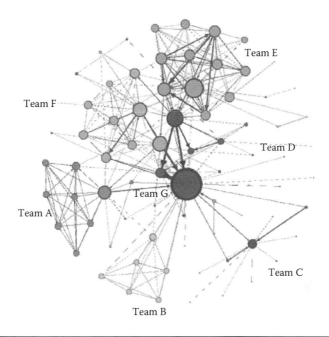

Figure 1.7   Aligning the structure of a product and the structure of the project.

interdependencies. Strongly connected elements of the system are displayed in the same shade of gray. It has been shown that a project structure typically mimics the system structure, as shown in this figure (Conway, 1968).

Although the architecture of a system and the structure of a project must be aligned, in a geographically distributed environment allocation of work across teams must also be planned by carefully examining dependencies among the elements of a system. Not only must strongly connected elements be allocated to a single colocated team, but also teams that have higher communication and coordination needs should not be located several time zones apart from each other.

Architecture keeps a system intellectually graspable and serves as a mechanism for communication, coordination, and control in geographically distributed projects.

1.5 Practice for Architecture-Centric Engineering

International Organization for Standardization/International Electrotechnical Commission (ISO/IEC) 42010 (ISO/IEC, 2011) is an international standard that provides a conceptual framework for incorporating architectural thinking into the development of a system. Figure 1.8 captures the key elements of this framework using UML (Unified Modeling Language). Every *system* exhibits an *architecture* that addresses the concerns and constraints of its *stakeholders* and its *environment*. These concerns and constraints become the basis for creating the architecture, which is documented using an *architecture description*.

Architecture serves as a basis for not only development of a system but also its maintenance and operation (ISO/IEC, 2011). As a system continues to evolve, each iteration of its development brings in new requirements from its stakeholders and its environment in response to changing business needs and technology. The architecture a system exhibits must have the flexibility to adapt to these changes.

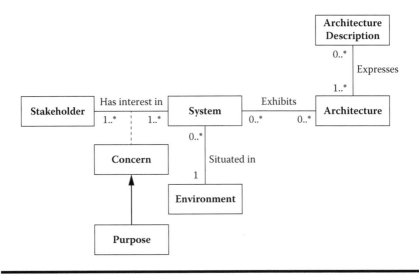

Figure 1.8 Conceptual framework for system and software architecture.

How central the focus on architecture is to a development process varies from one organization to another. Because architecture is associated with big design upfront (BDUF), many organizations following Agile processes place less emphasis on it than those that use a more plan-driven approach (Boehm and Turner, 2003). Advocates for Agile processes focus more on maximizing the flow of value to their customers and see architectural work as incurring cost and delaying delivery of value. However, plan-driven approaches suggest that ignoring architectural work in the short term can create conditions in the long term that make it difficult (or even impossible) to deliver value at a sustainable pace. This is more evident in large-scale projects; therefore, concepts such as Sprint 0 or architecture spike have been added for architectural envisioning to Agile development processes as well.

The practice of architecture-centric engineering (PACE) described in this book recognizes the significance of architecture in sustainable delivery of value throughout the life cycle of a product. PACE is more an approach to, rather than a process for, developing a system. It recognizes that a system must fulfill the purpose for which it was designed. The natural starting point for PACE, therefore, is a set of business goals or mission objectives of a system that define this purpose. By examining these goals, PACE essentially tries to answer the question, Why are we creating the given system? The answers to this and other questions become the basis for architecture envisioning for the given system.

Business goals or mission objectives capture the concerns of the stakeholders of a system and become the basis for its quality attribute requirements. When using two different versions of a system, you may find them functionally equivalent but may develop a preference for one over the other because of its quick response time, ease of use, ease of modification, or high reliability. Such characteristics of a system are called quality attributes and are the predominant forces that shape its architecture. PACE uses quality attributes as the basis for

HOW MUCH ARCHITECTURE IS ENOUGH?

Boehm and Turner (2004) analyzed several projects to understand the impact of architectural effort. As shown in Figure 1.9, they found that investment in architecture pays off in terms of reduced rework.

The sweet spot for how much architecture is essential can be determined by taking the total effort, architecture work and rework, into account. It turns out that the smaller projects require less investment in architecture up front when compared to larger projects. Smaller projects are characterized by a highly dynamic environment with rapid change and volatile requirements, whereas larger projects are characterized by more stable environments that may involve development of safety-critical products with limited uncertainty in their requirements.

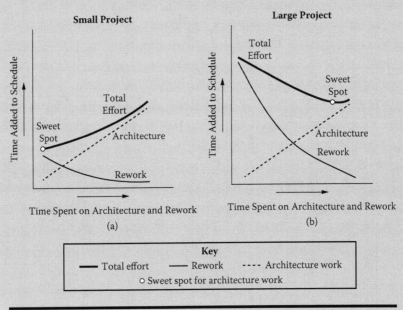

Figure 1.9   Architecture in (a) small-scale projects and (b) large-scale projects. (Adapted from B. Boehm and R. Turner, *Balancing Agility and Discipline: A Guide for the Perplexed*. Boston: Addison-Wesley, 2004.)

determining the most significant requirements that drive the development of the architecture of a given system. These requirements are expressed in the form of quality attribute scenarios that are not only unambiguous but also testable so they can also serve as test cases for validating the architecture.

Once the architecture is in place, it must be validated against stakeholder concerns to see if the system meets its business goals or mission objectives. If it does not, then one of two things must happen: either the mission objectives need to be changed or the architecture needs to be modified. Thus, architecture design is an iterative process that is carried out until all stakeholder concerns are addressed.

Once the stakeholder concerns have been addressed, the architecture becomes a baseline for detailed design. However, what is detailed design to one person could be architecture design to another. The requirements associated with baseline architecture can become drivers for defining the architecture of individual elements that make up this baseline. Thus, architecture design is a recursive process that is repeatedly applied to drive the design to a level with details sufficient to build the final system.

As the design process is applied recursively, the design decisions at each level serve as a starting point and context for the subsequent level. Architecture is, therefore, discovered within the confines of decisions already made, and it is managed by ensuring that future design decisions conform to the constraints set forth by the current decisions.

SUSTAINABLE DELIVERY OF VALUE

PACE makes sustainable delivery of value throughout the life cycle of a product possible by embracing several key practices:

- Stakeholder concerns express the purpose for why a system is being designed and are used for defining the business goals or mission objectives for the system under design.

- Mission objectives become the basis for defining quality attribute requirements, which serve as architectural drivers for the system.
- Unambiguous expression of quality attribute requirements as quality attribute scenarios that are also testable makes it easier to evaluate the architecture against stakeholder expectations.
- Architecture design is both an iterative and a recursive process. An iteration produces a baseline of an architecture that can recursively be refined to a level until there are sufficient details to realize the system. Design decisions made at each level become constraints for the next level of refinement.

Designing architecture is guided by patterns and tactics that are well-known solutions to commonly occurring design problems. In creating the architecture, attention must also be given to allocating functional responsibilities of the system to the different elements of the architecture. PACE uses the system context to determine all its actors, their use cases, and the corresponding functional responsibilities of the system. For ease of comprehension and consistency of expression of the use cases and functional responsibilities of the system, PACE uses the problem domain model. The model also plays a significant role in detailed design when functional responsibilities are ultimately allocated to the elements that realize the system.

1.6 Summary

Architecture plays a significant role in managing complexity, change, and distributed development of large-scale systems. It can be used for avoiding accidental complexity, adapting to change, and distributing development work commensurate with the communication, coordination, and collaboration needs of geographically distributed teams.

PACE is an approach described in this book for developing an architecture of a system that fulfills its business or mission objectives. The quality desired by these objectives is designed into the architecture using known architectural patterns and tactics, and the functionality is allocated to the elements of the architecture using the object-oriented analysis and design techniques.

1.7 Questions

1. Look at some of the other definitions of the term *architecture*. How are they similar? How are they different?
2. Conway's law (Conway, 1968) suggests that the structure or architecture of a system being developed and the organization developing it must be closely aligned. Can you see some of the difficulties that might arise as a result of their misalignment?
3. Do you think an existing organizational structure will constrain the architectures of future products that might be developed?
4. Give some examples of incidental and accidental complexity to illustrate the difference between the two.
5. Modularity, splitting a system into units or modules that are structurally independent of one another, is one way of creating flexible systems that can manage change effectively. What do you understand by this statement? Illustrate with an example.

References

D. Avery, The evolution of flight management systems, *IEEE Software* 28 (1), 11–13, 2011.

L. Bass, P. Clements, and R. Kazman, *Software Architecture in Practice,* second edition. Boston: Addison-Wesley, 2003.

B. Boehm and R. Turner, *Balancing Agility and Discipline: A Guide for the Perplexed.* Boston: Addison-Wesley, 2004.

F. Brooks, *The Mythical Man-Month.* Boston: Addison-Wesley, 1995.

S. Burger, O. Hummel, and M. Heinisch, Airbus cabin software, *IEEE Software* 30 (1), 21–25, 2013.

M. Conway, How do committees invent? *Datamation* 14 (5): 28–31, 1968.

R. De Neufville and S. Scholtes, *Flexibility in Engineering Design.* Cambridge, MA: MIT Press, 2011.

C. Hagen and J. Sorenson, Delivering military software affordably, *Defense AT&L* March–April 2013, pp. 30–34.

J. Herbsleb and R. Grinter, Splitting the organization and integrating the code: Conway's law revisited, *Proc. Int. Conf. Software Eng.* 1999, pp. 85–95.

R. Sangwan, M. Bass, N. Mullick, D. Paulish, and J. Kazmeier, *Global Software Development Handbook.* Boca Raton, FL: Auerbach, 2006.

R. Sangwan, K. Jablokow, M. Bass, and D. Paulish, Asynchronous collaboration: Achieving shared understanding beyond the first 100 meters. In *Proceedings of the ASEE Annual Conference and Exposition* (CD-ROM), June 18–21, 2006, Chicago (19 pages).

R. Sangwan and J. Ros, Architecture leadership and management in globally distributed software development. In *Proceedings of the First ACM Workshop on Leadership and Management in Software Architecture, International Conference of Software Engineering (ICSE),* May 11, 2008, Leipzig, Germany, pp. 17–22.

Chapter 2

Stakeholders and Their Business Goals

2.1 Introduction

Systems are created for a given purpose; therefore, it is critical to understand why we are building a given system. No matter how technically elegant, it is no good if a system's purpose is not clear and it fails to meet its intended objectives. These intended objectives are typically found in the strategic vision that prompted an organization to undertake the development of a system of interest.

Consider a company that sells building automation devices along with software applications that manage these devices. Software helps with the sale of these hardware devices but does not make any money. Over the years, the marketing and sales divisions have come to realize that the hardware is being commoditized and their profit margins on the sale of the hardware devices are shrinking. To sustain their business long term, the vice president (VP) of product development in consultation with the chief technology officer (CTO), decides to create a new building automation system that will be profitable. They wish to accomplish this by doing two things: reducing internal development costs and expanding the market. Internal development costs can be reduced by replacing several of the existing applications with a single building automation system. Market expansion can be achieved by entering new and emerging geographic markets and opening new sales channels in the form of value-added resellers (VARs). VARs sell the new building automation system under their own brand to support building automation hardware devices of many different manufacturers.

What are the intended objectives of the new building automation system? Do you think these objectives will significantly influence the architecture of the system of interest? Who are the stakeholders for these objectives?

2.2 Influence of Business Goals on the Architecture

The development of the new building automation system is prompted by the following business goals:

■ Reduction in internal development costs
■ Expansion of the product market
■ Partnering with VARs

Market expansion requires taking language, culture, and regulatory concerns of different markets into account. Partnering with VARs requires the new building automation system to work with hardware devices from many different manufacturers. Reduction in development costs may be achieved through consolidation of development and maintenance efforts for disparate software applications.

Certain business goals (such as market expansion and partnering with VARs) can give rise to requirements that can have a significant impact on the architecture of a system of interest; others (such as reduction in cost) may not (Sangwan and Neill, 2007). Depending on the context, there can be many such influential goals. Although in the case of a building automation system they were fairly explicitly stated, on other projects there may be a need for engaging stakeholders to elicit these goals. To help with this process, Table 2.1 lists typical categories of business goals that apply to projects undertaken by corporations.

Table 2.1   Categories of Such Business Goals

Goal Category	Goal Examples
Organization's growth and continuity	System promotes growth and continuity through • Long-term business sustenance • Market share increase • Product line creation and success • International sales
Meeting financial objectives	System meets financial objectives through • Revenue generation • Business process efficiency • Reduced training costs • Higher shareholder dividends • Employee bonuses
Meeting personal objectives	System meets personal objectives through • Enhanced reputation • Experience with new technologies • Experience with new development processes
Meeting responsibility to employees	System fulfills responsibilities to employee through • Opportunity for learning new development skills • Improved operator safety and reduced workload

(continued)

Table 2.1 Categories of Such Business Goals (continued)

Goal Category	Goal Examples
Meeting responsibility to country	System fulfills responsibilities to a country through • Compliance with export controls • Regulatory conformance
Meeting responsibility to society	System fulfills responsibilities to a society through • Compliance with laws and regulations, particularly those related to ethics, safety, security, and privacy • Green computing
Meeting responsibility to shareholders	System fulfills responsibilities to shareholders through • Adherence to Sarbanes-Oxley Act • Liability protection
Managing market position	System helps manage market position through • Retention or increase in market share • Intellectual property protection
Improving business processes	By improving business processes, a system creates opportunities for • New markets and products • Better customer support
Managing product's quality and reputation	System helps • Improve branding • Reduce recalls • Support certain types of users • Improve quality and testing support and strategies

Source: Based on P. Clements and L. Bass, *IEEE Software* pp. 38–46, November/December 2010.

There is some overlap within these categories, but the intent is not for them to be mutually exclusive. They are simply there to serve as a vehicle for facilitating the elicitation of business goals from the stakeholders of a system under design.

When eliciting these objectives, care must be undertaken to engage all stakeholders from the organizations involved in development, acquisition and operation of a system that can include any group or individual affected by or in some way accountable for the outcome of an undertaking. Because arriving at an agreed-to set of these objectives can be a long and arduous process, it is important also to know who the primary stakeholders are that exert a greater degree of influence on a project and have the authority to help resolve conflicts when they arise (National Aeronautics and Space Administration [NASA], 2007).

> Business goals define the purpose for which a system is being created. The architecture is designed to fulfill that purpose.

2.3 Representing Business Goals

A business goal scenario is a six-part structure that makes a business goal concrete and actionable (Clements and Bass, 2010):

- *Goal subject* (GS): the stakeholder who owns the goal
- *Goal object* (GO): the entity to which the goal applies
- *Environment* (E): the context for the goal
- *Goal* (G): the goal itself
- *Goal measure* (GM): the success criteria for a goal
- *Pedigree* (P): background that includes identification of the person or artifact providing the goal statement, how much confidence there is in the goal, the goal's expected volatility over time, and the goal's value

We show the business goals for the building automation system in Table 2.2. Initially, the business goals may be

Table 2.2 Business Goals for the Building Automation System

Goal Category	Informal Business Goal Statement	Business Goal Scenario
Organization's growth and continuity	We need to reduce the internal development cost by replacing several applications with a single building automation system.	• **GS:** CTO
		• **GO:** development organization
		• **E:** proliferation of applications that incur development and maintenance costs but do not generate revenue
		• **G:** reduce internal development costs
		• **GM:** replace applications with a single building automation system that can be sold for profit
		• **P:** marketing and sales reports show commoditization of hardware and declining profit margins, putting long-term sustenance of the business in jeopardy
Meeting financial objective	We need to expand by entering new and emerging geographic markets.	• **GS:** VP of product development
		• **GO:** system
		• **E:** organization needs to generate additional revenues
		• **G:** bring system to new and emerging geographic market
		• **GM:** add five new customers during the first year

(continued)

Table 2.2   Business Goals for the Building Automation System (cont.)

Goal Category	Informal Business Goal Statement	Business Goal Scenario
		• **P:** marketing and sales reports show commoditization of hardware and declining profit margins, putting long-term sustenance of the business in jeopardy
Meeting financial objective	We need to open up new sales channels through partnership with VARs.	• **GS:** VP of product development
		• **GO:** system
		• **E:** organization needs to generate additional revenue
		• **G:** open up new sales channel in the form of VARs
		• **GM:** increase revenue by 20% on an annual basis
		• **P:** Marketing and sales reports show commoditization of hardware and declining profit margins, putting long-term sustenance of the business in jeopardy

captured informally as shown in the middle column and later expressed as detailed scenarios shown in the leftmost column in Table 2.2.

A business goal scenario makes a business goal concrete and actionable.

2.4 Refining Business Goals

One observation to make is that business goals can be fairly broad, such as a company's desire to be profitable, and may need refining. As also pointed out previously, we must carefully examine the business goals and see if they give rise to requirements that have an influence on the architecture of the system under development. For example, profitability may be achieved by making changes to the business processes of a company and therefore may not have anything to do with the architecture of the system to be developed.

Table 2.3 shows the business goals and their relevance to the building automation system introduced in the previous section. Business goals are synonymous with *mission objectives,* a term typically used by the Department of Defense (DoD) and NASA when creating architectures of weapon and space systems (NASA, 2007). Just like business goals are refined to determine how they apply to the system under development, mission objectives are refined into *operational objectives,* some of which would be relevant to the product being engineered. The relevant operational objectives are sometimes called *engineering objectives.*

Notice that the reduction in internal development costs can be achieved by aggregating the development and maintenance

Table 2.3 Business Goals and Engineering Objectives for the Building Automation System

Business Goal (Mission Objective)	Goal Refinement (Engineering Objective)
Expand by entering new and emerging geographic markets	Support international languages
	Comply with regulations that have an impact on life-critical systems such as fire alarms
Open new sales channels in the form of VARs	Support hardware devices from different manufacturers
	Support conversions of nonstandard units used by the different hardware devices

functions of several small applications into one project for the building automation system. It has more to do with changes to the organizational structure than the architecture of the system to be developed and therefore is not included in Table 2.3.

> Engineering objectives refine business goals/mission objectives to show their relevance to the system under design.

2.5 Translating Engineering Objectives into Architectural Requirements

Another important observation to make is that business goals relevant to a system correspond to quality attributes the end system must exhibit. These quality attributes become the basis for translating the business goals and their associated engineering objectives into architectural requirements. For instance, a business goal to maintain market reputation can correspond to creating products that are highly usable and reliable and perform well. Can you think of some of the quality attributes that correspond to the business goals of the aforementioned building automation system?

AUTOMOTIVE SYSTEMS

With crowded highways, traffic congestion, and rising fuel prices, consumers are looking for cars that are not only safe but also less dependent on gasoline. The worldwide emissions legislation for automobiles is also requiring a reduction in fuel consumption and emissions to reduce greenhouse gases that cause environmental pollution. These factors that have become the cornerstone for innovation in the automotive industry must be largely achieved with the help of electronics and software. They are the business goals or mission objectives that are driving the manufacturing of modern-day automobiles. Some significant quality attributes associated with these business goals are the following:

- **Reliability**: Automotive software running on a complex network of electronic control units (ECUs) must be exceptionally reliable.
- **Safety**: Functions such as antilock braking and electronic stability control require fail-safe operation.
- **Performance**: Safety-critical scenarios require a real-time response in the range of micro- to milliseconds.
- **Efficiency**: Reduction in fuel consumption and emissions requires improved powertrain and propulsion technologies.

As shown in Figure 2.1, a high-end vehicle today has upward of 80 ECUs that communicate over a complex in-vehicle network that runs over 100 million lines of code to satisfy these quality attribute requirements, and you never hear of a blue screen of death or the need to reboot when operating a vehicle; it is simply not an option (Mossinger, 2010).

To support a multitude of hardware devices and consider different languages and cultures, the building automation system must be able to accommodate these changes easily (a modifiability requirement). To support different regulations in different geographic markets, the system must respond to life-threatening events in a timely manner (a performance requirement). It is critical that the business goals and their implied quality concerns be fully understood because they are the very reason a system is created. They will play a significant role in the creation of architecture for a given system. If ignored or not fully understood, you risk creating a system that may not be fit for its intended purpose. We list some predominant quality attributes (Bass et al., 2013) and their description in Table 2.4.

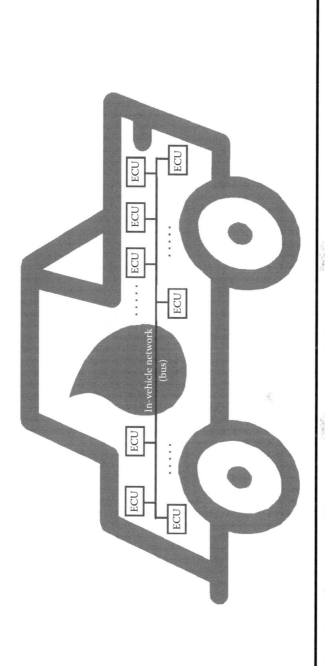

Figure 2.1 A modern automobile with a complex network of ECUs.

Table 2.4 Quality Attributes and Their Descriptions

Quality Attribute	Description
Availability	Concerned with preventing, detecting, and recovering from system failures
Interoperability	Concerned with meaningful exchange of information among two or more systems
Modifiability	Concerned with the impact of making a change to a system
Performance	Concerned with latency and throughput measured as the length of time it takes for a system to respond to an event (latency) and how many events a system can respond to in a given period of time (throughput)
Security	Concerned with a system's ability to resist unauthorized usage while providing its services to legitimate users with the following guarantees: • *Nonrepudiation*: A transaction cannot be denied by any of the parties to it. • *Confidentiality*: Data or services are protected from unauthorized access. • *Integrity*: Data or services are delivered as intended. • *Assurance*: Parties to a transaction are who they claim to be. • *Availability*: The system is available for legitimate use. • *Auditing*: The system tracks all of its own activities.
Testability	Concerned with the ease with which a system can be made to demonstrate its faults through testing
Usability	Concerned with the ease with which a user can perform a given task and the type of support a system provides for it

Table 2.5 Business Goals, Engineering Objectives, and Their Corresponding Quality Attributes

Business Goal (Mission Objective)	Goal Refinement (Engineering Objective)	Quality Attribute
Expand by entering new and emerging geographic markets	Support international languages	Modifiability
	Comply with regulations that have an impact on life-critical systems such as fire alarms	Performance
Open new sales channels in the form of VARs	Support hardware devices from different manufacturers	Modifiability
	Support conversions of nonstandard units used by the different hardware devices	Modifiability

Table 2.5 lists the business goals, their refinement (showing relevance to the system under consideration), and their corresponding quality attributes.

Quality attribute requirements must provide sufficient detail to be truly useful. As an illustration, imagine you are the architect for the building automation system and suppose you have a requirement that the building automation system must be modifiable with respect to the hardware devices it must support from different manufacturers. Do you think the requirement is sufficient in its given form?

It may not be sufficient to say that a system must be modifiable. Any system is modifiable with respect to something, and a system can be modified with respect to any aspect given enough time and money. The question relates to its modifiability with respect to what, when, and the amount of effort needed to do so. The given requirement for the building automation system clearly states the what, but not when and how much effort. It may be more useful to state the requirement as follows:

> A field engineer is able to integrate a new field device
> into the system at runtime, and the system continues
> to operate with no downtime or side effects.

Now, it is clear that a new field device (what) has to be added
at runtime (when) by a nonprogrammer (how much effort).

This form of expressing a requirement is called a *quality
attribute scenario* (Bass et al., 2003). Stated this way, the quality attribute requirement is more concrete. Architectural design
decisions can be made to satisfy the requirement, and the
system can be tested to see if it meets this requirement.

A quality attribute scenario is a requirement with six parts:
a stimulus, the source of stimulus, an environment, an artifact,
a response, and a response measure. These six parts for the
scenario just described are as follows:

Stimulus—integrate a new field device
Source of stimulus—field engineer
Environment—runtime
Artifact—the system
Response—system continues to operate
Response measure—with no downtime or side effects

So, the stimulus is something that incites a system, the source of
stimulus is something that generates the stimulus, the environment is the condition under which the stimulus is generated,
the artifact is the part of the system that is affected by the system, response is the reaction of the system to the stimulus, and
response measure is a way for measuring the system's reaction.

> Engineering objectives have implied quality concerns that
> are elaborated into architectural requirements expressed
> as quality attribute scenarios.

2.6 Prioritizing Architectural Requirements

Because they are the drivers for the architectural decisions, the first task is to determine the important quality attribute requirements for a given system. You may have observed one important aspect of quality attribute requirements is that they typically conflict with each other. For instance, trying to increase the security within a system through the use of encryption may have a negative impact on its response time. For any given system, therefore, you can only have a handful of such requirements; otherwise, they can create conditions that make conflicts challenging to resolve.

How might you go about discovering this set of most important quality attribute requirements for a system under consideration?

One approach may be to consider a standard taxonomy of quality attributes such as International Organization for Standardization (ISO) 9126. You may examine each quality attribute in this taxonomy and see how it applies to your system. As you go down the list, it will be hard to find any quality attribute that does not apply. After all, who will say that their system does not need to be secure, reliable, usable, testable, scalable, and so on? So, this approach does not seem practical.

Another way you could approach this problem is to examine each function to be supported by the system and ask the stakeholder for that function to describe some quality attributes associated with that function. Again, it would be hard for any stakeholder to resist saying the function desires should be high performing, secure, reliable, usable, and so on. This approach seems even worse than the first one because the quality attribute requirements in this case will be some multiple of the total number of functional requirements. Clearly, this number is much more than a handful.

A pragmatic approach is to elicit quality attribute scenarios from stakeholders that directly relate to the business goals for a system. Because there are only a handful of business goals, the same will apply to the quality attribute scenarios. More significantly, these would be the most important scenarios to consider if one is to create a system that fulfills its intended purpose (Ozkaya et al., 2008).

Quality Attribute Workshop (Barbacci et al., 2003) is an architecture-centric method that can be used for eliciting quality attribute requirements from the stakeholders of a given system. The goal of this method is to establish a prioritized set of quality attribute requirements in the form of quality attribute scenarios that are mapped to the business goals. Clearly, it is important that these goals are known before the workshop can be conducted even if they are initially general and will need subsequent refinement.

Table 2.6 shows a prioritized set of scenarios derived from the business goals for the building automation system.

The priorities are based on how business critical a quality attribute requirement is. High (H) means the customers would not consider the product if that requirement is not satisfied, medium (M) means the product would not be competitive in the absence of that requirement, and low (L) means it is something nice to have.

> Quality attribute requirements often conflict with each other, leading to complex trade-off situations. Hence, it is important to prioritize these requirements and only consider half a dozen or fewer most significant ones.

2.7 Summary

The ISO/IEC (International Electrotechnical Commission) 42010 standard was introduced briefly in Chapter 1. This chapter explored the shaded portion of this conceptual framework shown in Figure 2.2.

Table 2.6   Quality Attributes and Scenarios Derived from Engineering Objectives

Engineering Objective	Quality Attribute	Quality Attribute Scenario	Priority
Support hardware devices from many different manufacturers	Modifiability	A field engineer is able to integrate a new field device into the system at runtime, and the system continues to operate with no downtime or side effects.	H
Support conversions of nonstandard units used by the different devices	Modifiability	A system administrator configures the system at runtime to handle the units from a newly plugged in field device, and the system continues to operate with no downtime or side effects.	H
Support international languages	Modifiability	A developer is able to package a version of the system with new language support in 80 person-hours.	M
Comply with regulations	Performance	A life-critical alarm should be reported to the concerned users within 3 s of the occurrence of the event that generated the alarm.	H

Note: H, high; L, low; M, medium.

It is important to know who are the key stakeholders with a vested interest in the system under design. The stakeholders have concerns that define the purpose for which the system is being created. These concerns are expressed as business goals or mission objectives and can have a significant impact on the

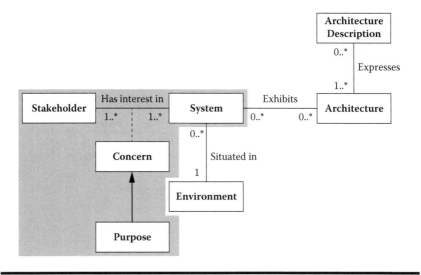

Figure 2.2 Conceptual framework for system and software architecture.

architecture of a system. Therefore, they must be used as a starting point for determining architecturally significant requirements. Because not all business goals may be relevant, they must first be refined into engineering objectives that relate them to the system under design. The engineering objectives are then refined into quality attribute scenarios. Because these scenarios may conflict with each other, they need to be prioritized, and only a handful of the most significant ones should be considered to avoid complex trade-off situations.

2.8 Questions

1. The Pareto principle states that, for many events, roughly 80% of the effects come from 20% of the causes. If we were to apply this 80–20 rule to the systems development world, we can argue that 20% of the highest-priority features can satisfy 80% of the business needs. In other words, 80% of the features seldom add value to the system under consideration (see Johnson, 2002). How does a

clear focus on business goals or mission objectives when creating a system help avoid non-value-added work?

2. Why is it important to refine business goals into engineering objectives?

3. Life-critical systems typically have a requirement to operate continuously. Refine the appropriate business goal for the building automation system into an engineering objective for continuous operation.

4. Map the engineering objective in problem 3 to a quality attribute and elaborate it using a quality attribute scenario. Identify the six parts of the scenario.

5. Look at the automotive systems case study presented in this chapter. List all the important business goals, refine them into engineering objectives, identify the implied quality concerns, and develop at least one quality attribute scenario corresponding to these concerns.

References

M. Barbacci, R. Ellison, T. Lattanze, J. Stafford, C. Weinstock, and W. Wood, *Quality Attribute Workshops (QAWs),* third edition (CMU/SEI-2003-TR-016). Pittsburgh, PA: Software Engineering Institute, Carnegie Mellon University, 2003.

L. Bass, P. Clements, and R. Kazman, *Software Architecture in Practice,* third edition. Boston: Addison-Wesley, 2013.

P. Clements and L. Bass, The business goals viewpoint, *IEEE Software,* November/December, 2010, pp. 38–46.

International Organization for Standarization. ISO/IEC Standard 9126: Software Engineering — Product Quality, part 1. 2001.

ISO/IEC/IEEE Systems and software engineering—Architecture description, ISO/IEC/IEEE 42010:2011(E) (Revision of ISO/IEC 42010:2007 and IEEE Std 1471-2000), pp. 1–46, 2011.

J. Johnson, *Chaos Report.* Boston: Standish Group, 2002.

J. Mossinger, Software in automotive systems, *IEEE Software* March/April 2010, pp. 92–94.

National Aeronautics and Space Administration (NASA), *Systems Engineering Handbook* (NASA/SP-2007-6105 Rev1). Washington, DC: NASA Headquarters, December 2007.

I. Ozkaya, L. Bass, R. Sangwan, and R. Nord, Making practical use of quality attribute information, *IEEE Software*, March/April 2008, pp. 25–33.

R. Sangwan and C. Neill, How business goals drive architectural design, *IEEE Computer*, August 2007, pp. 101–103.

Chapter 3

Establishing Broad Functional Understanding

3.1 Introduction

In the previous chapter, we learned how to capture, in a concrete and actionable form, expectations stakeholders have for the system under consideration. These expectations are tied to higher-level business goals or mission objectives that define the purpose for which a system is being created. A system, however, exists within a context or operates within an external environment. Understanding a system's operational aspects within its environment is also extremely important. It helps clearly delineate where a system's boundary is, what elements in its external environment a system must interact with, and what those interactions are.

In military and government projects, the term *concept of operations* or ConOps is used to establish a broad operational understanding of the system in its environment. This broad understanding includes a system context, system use cases or operational scenarios, an end-to-end operational view of the system, and any design constraints along with a rationale for those constraints (American National Standards Institute/ American Institute of Aeronautics and Astronautics [ANSI/ AIAA], 2012; National Aeronautics and Space Administration [NASA], 2007). We discuss each of these elements in the sections that follow.

> ConOps is an operational view of the system from the perspective of its users, and it gives a broad understanding of the capabilities a system must deliver to fulfill its mission objectives.

3.2 System Context

A system context shows the system as a black box along with all external entities (human and nonhuman) with which it must interact in its operational environment. Figure 3.1 shows a system context diagram for the building automation system we introduced in the previous chapter.

The diagram in Figure 3.1 shows the building automation system with all its actors, a term used in Unified Modeling Language (UML) to describe the external entities with which a system must interact. For the building automation system, these include field devices that report operating conditions of a building, a facility manager who monitors these conditions, a field engineer who deploys and maintains the field devices, a system administrator who administers the system and its users, and a public safety system that is notified under conditions that threaten the safety and security of residents in the building being monitored.

> System context captures all external entities with which a system must interact in its operational environment.

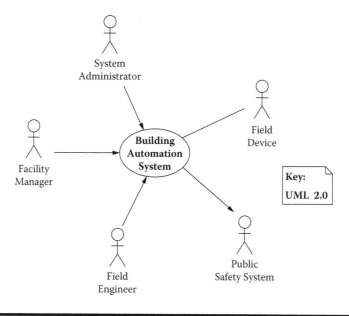

Figure 3.1   System context for the building automation system.

3.3 System Use Cases

Functional capabilities or operational scenarios of a system can be represented with use cases based on how the external actors shown in the context diagram in Figure 3.1 intend to use the system. For instance, the field engineer intends to manage field devices and dynamically reconfigure them. The facilities manager intends to manage alarms generated by field devices that monitor a building. Alarms related to events that could cause loss of life also result in notifications to the public safety system. The system administrator intends to manage the users of the building automation system.

Some of the use cases related to the goals of the actors of the building automation system are shown in Figure 3.2. These use cases can be grouped by product features they realize (shown as notes on the right of Figure 3.2) and provide a broad functional context of the system under development.

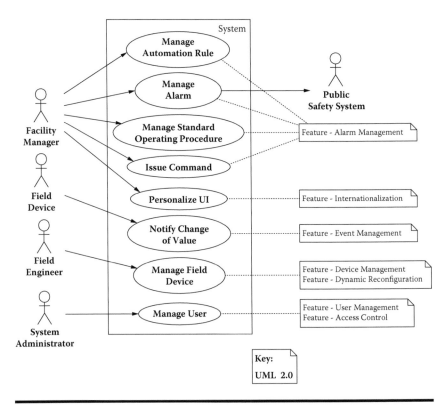

Figure 3.2   Use cases for the building automation system.

The use cases that implement the *Alarm Management* and *Internationalization* features are designed for the facilities manager. These include *Manage Automation Rules* for creating, updating, or deleting rules for automating building functions; *Manage Alarm* for responding to alarms generated by the system in response to a life-critical situation; *Manage Standard Operating Procedures* for creating, updating, or deleting standard operating procedures for handling policies related to monitoring the health of a building; *Issue Command* for commanding a field device (to reset itself, for example); and *Personalize UI* for customizing the user interface to a particular language and locale.

The *Event Management* feature includes a single use case, *Notify Change of Value,* used by a field device to notify change in any of its property values (temperature value change in a thermostat, for example). The features *Field Device Management* and *Dynamic Reconfiguration* are implemented through the use case *Manage Field Device* used by a field engineer. Finally, *User Management* and *Access Control* are provided through the use case *Manage User* used by a system administrator.

Although these are brief descriptions of the use cases, the elaboration of each use case can be fairly detailed (Larman, 2005). For illustration, the use case *Issue Command* is fully elaborated in Figure 3.3. The use case is given a unique identifier (Use Case 001) for cross-referencing. The elaboration

---

Use Case 001: Issue Command

> The user selects a field device property to modify, specifies the new value of the property and sees the change reflected.

Preconditions

> 1. The user is authenticated
> 2. The user has authorized access to the field device for which the property is to be modified

Postconditions

> 1. The selected field device property is modified

Main Success Scenario

> 1. User selects a field device
> 2. System displays the properties (to which the user has read access) of the selected field device
> 3. System verifies and enables "Issue Command" for those properties for which the user has write access
> 4. User selects "Issue command" for the property to be modified
> 5. System prompts the user for the new value of the property
> 6. User inputs a new value and confirms
> 7. System transmits the command to the field device
> 8. Field device sends a change of value notification, confirming the success of the command
> 9. System displays the new property value and confirms command success

Extension 001.3a
3a - User has no write access to the selected property:

> 1. "Issue Command" is disabled

Extension 001.6a
6a - Erroneous value entered:

> 1. System notifies the user of the incorrect value.
> 2. System does not send the command to the field device.

Extension 001.8a
8a - Command fails at the filed device level:

> 1. System notifies the user of command failure and reason of failure.

Figure 3.3 A fully elaborated use case.

consists of a brief description of the purpose of a use case followed by its preconditions, postconditions, the main success scenario, and extension points. Preconditions are things that must be true before the use case can be executed by the system; postconditions are things that must be true after the use case is executed by the system; a main success scenario consists of interactions between the user and the system that led to the desired outcome; and extension points are exceptions that might occur during the execution of the main success scenario. Extension points are numbered so it is clear which step of the given use case's main success scenario caused the exception. For instance, Extension 001.6a is an exception that occurred at step 6 of the main success scenario for Use Case 001. If other exceptions were to occur at the same step, they would be labeled Extension 001.6b, Extension 001.6c, and so on.

The elaborated use cases can be used to determine the responsibilities (see responsibility-driven design by Wirfs-Brock and McKean, 2003) or operational capabilities of a given system. For instance, the following operational capabilities for the building automation system can be derived from the use cases shown in Figure 3.2:

1. Send user commands to a field device
2. Receive events from a field device
3. Perform semantic translation for field device data
4. Configure a field device
5. Route data to a field device
6. Evaluate and execute automation rules
7. Send automation commands
8. Generate alarm notifications
9. Display device data
10. Capture/relay user commands
11. Display alarm notifications
12. Edit/create automation rules
13. Retrieve data from a field device

14. Store field device configuration
15. Propagate change-of-value notifications
16. Authenticate and authorize users
17. Persist automation rules, user preferences, alarms

> Use cases are interactions initiated by users or external systems (referred to as actors). Use cases are goal oriented; in other words, at the end of a use case the actor initiating the given use case walks away from the system with something of value. They can be used for determining operational capabilities a system must support.

3.4 Domain Model

The descriptions of use cases introduce some significant concepts from the building automation domain, such as field device, field device properties, rules, alarms, and standard operating procedures. These concepts are significant because they represent the information/entities that are created, destroyed, associated, and used in various ways within a use case to achieve something of value. A conceptual or domain model is used to capture these significant terms and the relationships among them (Larman, 2005). Figure 3.4 shows this model for the building automation system using a UML class diagram. The convention of reading relationships shown in the diagram is top to bottom and left to right.

The domain model establishes a standard vocabulary of concepts that can be used when elaborating use cases and provides a succinct description from the problem domain. When examining the model in Figure 3.4, one can describe the building automation domain with relative ease. A *field engineer* manages *field devices* that monitor the internal operations of a building. The field devices can generate *alarms*

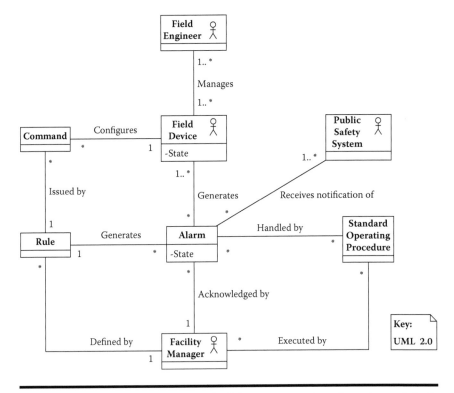

Figure 3.4   Key concepts of the building automation domain and their relationship.

under conditions that create a safety hazard. The alarms are monitored by *facility managers,* who follow *standard operating procedures* when handling these alarms. Facility managers can also define *rules* that can automatically evaluate hazardous conditions and generate alarms or can automatically issue *commands* to field devices to take certain actions.

Such a succinct description not only can be a great aid in establishing shared understanding of the problem domain but also can vastly improve the quality of system requirements because the standard vocabulary introduces a level of consistency, making requirements less subject to different interpretations. There are some additional applications of a domain model that are discussed in further chapters.

A domain model can be used to capture significant concepts and their relationships, creating a shared understanding of a problem domain, and can be used to establish a standard vocabulary that can be used to elaborate use cases consistently.

3.5 An End-to-End Operational View

The overall goal of the building automation system is to monitor the locations within the buildings (using field devices) for normal building operations. These devices periodically report back the conditions, and the facility managers can monitor these conditions on their workstations. Conditions that may threaten the safety and security of the building residents are raised as alarms so that prompt action can be taken. Figure 3.5 shows this end-to-end operational view of the system.

An end-to-end operational view provides a good conceptual understanding of the capabilities of a system. One can formulate an idea of what the system can do if it were available today. Operational scenarios or timelines based on this view can further provide details on time-sequenced order of major events the system must handle (NASA, 2007). Figure 3.6 shows one such operational scenario in which many field devices concurrently transmit events to the building automation system. The building automation system can process these events (and possibly request more data from the devices) to determine if there is a safety hazard, in which case an alarm is generated. The alarm needs to be displayed on a facility manager's workstation, and a notification also needs to be sent to the public safety system.

Such end-to-end operational scenarios can be useful in determining system configurations and operational activities necessary to achieve the engineering objectives. For instance,

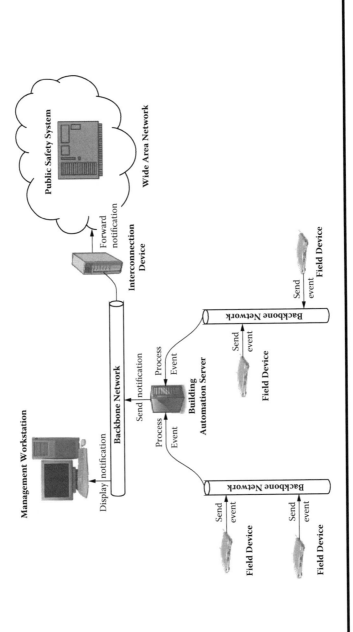

Figure 3.5 Operational view of the building automation system.

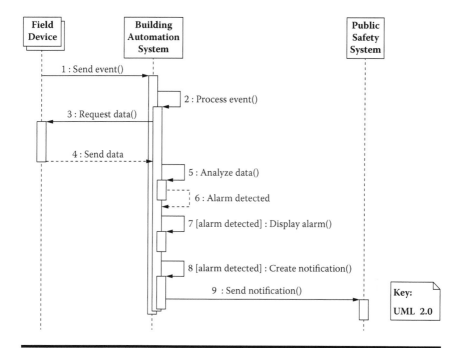

Figure 3.6 Operational scenario showing flow of major events.

the execution snapshot in Figure 3.6 can be used as a basis for creating an initial performance model to obtain an estimate of the processing requirements of the building automation system to handle event volumes, event arrival rates, and timely dissemination of alarm notifications related to safety-critical situations.

Recall that one of the engineering objectives was that, in the event that data from the field devices begin to indicate the possibility of an alarm, the facilities managers and possibly the public safety system need to know about this possibility within 3 s of its occurrence. Under normal operating conditions, a single field device can generate 1 event/second in the worst case.

Table 3.1 specifies the computing requirements for each of these processing steps for a single event received from a field device. Work units represent central processing unit (CPU) consumption, the range being 1 to 5 with 1 representing CPU consumption for a simple task and 5 representing CPU

Table 3.1 Computing Requirements

Processing Step	Work Unit	Network Message
Send event (field device)	0	1
Process event (building automation system)	5	4

consumption for a complex task. Network messages represent outbound messages from a processing step.

A sending event step does not require any processing (hence, a work unit value of 0) but merely transmission of data via a single network message from the field device to the building automation system. However, the process event step can be quite involved. First, the building automation system has to do semantic translation of event data received from a field device and evaluate it using a rule. The data may indicate an alarm condition and require additional data requests from other devices. For example, an event from a fire alarm may prompt a request for data from smoke detectors and sprinkler systems. We assume, in the worst case, an alarm condition leads to requests for data from two additional field devices (therefore, two data request messages to the devices). Finally, all the aggregated data need to be analyzed, possibly running several rules to determine if a situation is safety critical. So, a process event is a rather complex task; therefore, it is assigned a work unit value of 5. The total number of network messages is 4, assuming 2 messages for additional field devices mentioned previously and 1 message each to the facility manager workstation and the public safety system.

Table 3.2 shows the computing resources with their service times and the associated processing overhead. In the left section of the table, the names of computing devices in a typical server are in the leftmost column, the units of service provided by these devices in the second column, and the service time for each unit of service in the third column. For example, a CPU executes 1,000 instructions in 10 μs on a typical server that has

Table 3.2   Computer Resources and Processing Overhead

Computer Resources			Processing Overhead	
Device	*Unit of Service*	*Service Time/ Unit of Service*	*Work Unit*	*Network Message*
CPU	Thousand instructions	0.00001 s	20	10
Disk	Physical I/O	0.02 s	0	2
Network	Message	0.01 s	0	1

a single CPU. The values in the right section of the table define processing overhead associated with the computing needs (work units and network messages) of the building automation system. For example, reading down the left processing overhead column, a single work unit results in the execution of 20,000 (20K) CPU instructions, no disk inputs/outputs (I/Os) or network messages. Reading down the right processing overhead column, a network message requires 10K CPU instructions, two physical I/Os, and transmission of one network message.

Tables 3.1 and 3.2 can be used for calculating total computer resource requirements for our operational scenario. We show this in Table 3.3.

The best-case elapsed time for the alarm detection scenario, therefore, is $(150 \times 0.00001) + (10 \times 0.02) + (5 \times 0.01) = 0.15150$ s, which is well within our 3-s latency constraint for an alarm. This, however, is a ballpark estimate and does not take into account any network latency, unavailability, and

Table 3.3   Total Computer Resource Requirements for Alarm Detection

Processing Step	*CPU Instructions (K)*	*Physical I/O*	*Network Messages*
Send event	10	2	1
Process event	140	8	4
Total	150	10	5

queuing delays. We can continue to refine this model and its associated estimate as the design of the system matures and there is a better understanding of the event arrival rates and the topology of the system.

We can also try to gain a ballpark understanding of how we can handle the system load. Assume the building automation system is deployed on a 128-processor high-end server. The throughput of the server will be (128/0.15150) = approximately 845 events per second. To attain job flow balance, however, the system will have to be fast enough to handle the alarm events; thus, the throughput must be equal to the arrival rate of these events. Suppose the largest customer for the building automation system has 500,000 field devices on the premises. Assuming only 10% of these field devices are related to the safety-critical aspect of the building automation system (e.g., fire alarms, intrusion, etc.), we would need a cluster of (50,000/845) = approximately 60 servers just to handle the load created by these devices.

Although this is a ballpark analysis, such analysis can be valuable in understanding the flexibility that needs to be designed into the system such that it can be configured for handling the latency and throughput needs of its operational scenario. In case of the given scenario, the system would have to be designed with flexibility such that it can be deployed on a single low-end server for a smaller customer or on a cluster of high-end servers for a larger customer. Essentially, it should scale gracefully to accommodate varying degrees of load it needs to handle.

3.6 Constraints

There are additional stakeholder concerns that have to be considered as they may constrain how a system is designed. For instance, it may be the case that the system under consideration has to be developed using a certain technology

Table 3.4   Constraints for Designing the Building Automation System

Category	Factor	Description
Organization	Expertise	The development organization has a strong background in development on the Microsoft platform.
Technology	Cost	The system must reuse the company's existing rules engine.
Product	Variability	The system must handle a wide range of disparate field devices and configurations, ranging from 500 to 500,000 such devices at customer sites.

suite, such as the Microsoft platform and the Oracle DBMS. This a priori choice of technology will limit the ability of an architect to make design decisions, such as how the system is partitioned into tiers, the communication mechanisms across these tiers, and the strategies for security, fault, and transaction management. Table 3.4 enumerates a few such factors for the building automation system.

These factors have been broadly categorized into those that result from the capabilities or limitations of an organization to work with a particular technology suite or type of system, a mandate that the project use a particular technology to promote strategic reuse or contain cost, and the characteristics of the product, such as the types of customers it needs to serve or the regulatory concerns it must address.

Constraints are decisions that have been made to limit the freedom an architect has with respect to those decisions, therefore constraining how the system must be designed.

3.7 Summary

This chapter explored the shaded portion of ISO/IEC 42010 conceptual framework for system and software architecture shown in Figure 3.7.

ConOps can be used to understand the operational environment of a system under design. ConOps establish this operational understanding through the use of system context, system use cases or operational scenarios, domain model, an end-to-end operational view of the system, and design constraints. System context captures all external entities or actors a system must interact with in its operational environment. The intent of how these actors want to use the system becomes the basis for defining use cases and, finally, the responsibilities the system must fulfill. The use cases and responsibilities are expressed using a standard vocabulary established through the model of the problem domain. An end-to-end operational view of the system uses use case scenarios to capture a time-ordered sequence of major events a system must handle. This helps understand system configurations and operational activities necessary for achieving the capabilities a system must

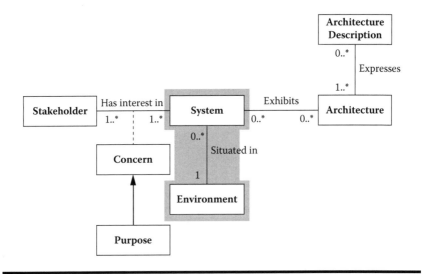

Figure 3.7   Conceptual framework for system and software architecture.

have. Constraints can limit the freedom of an architect in how the system must be designed to achieve these operational capabilities.

3.8 Questions

1. Pick one use case from Figure 3.2 and elaborate using the format shown in Figure 3.3. Make sure you express the use case using the vocabulary of the problem domain as established in Figure 3.4.
2. From the use case elaborated in problem 1, identify key operational capabilities that must be assigned to the building automation system.
3. During elaboration of the use case, were there any significant terms or concepts that you found were not represented in the domain model shown in the chapter? If so, augment the domain model by adding these concepts.
4. Consider an end-to-end scenario in which a user wants to change a property value (such as the temperature setting of a thermostat) of a device. Using the technique described in Section 3.5, model and estimate the elapsed time between the moment the user issues the command and the moment when the property value change takes effect.

References

American National Standards Institute/American Institute of Aeronautics and Astronautics (ANSI/AIAA), *ANSI/AIAA G-043A-2012 Guide for the Preparation of Operational Concept Documents*. Reston, VA: AIAA, 2012.

C. Larman, *Applying UML and Patterns*, third edition. Upper Saddle River, NJ: Prentice-Hall, 2005.

National Aeronautics and Space Administration (NASA), *Systems Engineering Handbook* (NASA/SP-2007-6105 Rev1). Washington, DC: NASA Headquarters, December 2007.

R. Wirfs-Brock and A. McKean, *Object Design: Roles, Responsibilities, and Collaborations*. Boston: Addison-Wesley, 2003.

Chapter 4

Getting Ready for Designing the Architecture

4.1 Introduction

The previous chapters focused on obtaining a good set of requirements that clearly establish the business need for the system under consideration and the qualities and functional capabilities this system must have to fulfill this need within the confines of a given a set of constraints. Of these requirements, the quality attribute scenarios developed from the business goals or mission objectives play the most significant role in the development of the architecture. However, operational scenarios for functional capabilities and the constraints may also have implied quality concerns that can have an impact on the architecture of a system and should be taken into account as well. We call all such requirements (quality attribute requirements and functional capabilities and constraints with implied quality concerns) that influence the architecture architecturally significant requirements (ASRs).

Although one may wish to take into consideration all ASRs when designing the architecture, these requirements typically conflict with each other and create complex trade-off situations that require an architect to look for a design that can try to achieve some balance among the opposing forces. For instance, when designing an architecture that makes the system modular for ease of making a change, one may introduce layers of indirection so that many modules may have to interact with each other to perform a given function. These interactions may therefore introduce latency that has an impact on how quickly the function can be performed. Thus, the two qualities (modifiability for ease of change and performance for quick response time) conflict with each other, and the balance may lie in finding the happy middle ground where the architect trades ease of making a change to achieve a faster response time.

TRADING COST AND EFFECTIVENESS

Cost is a constraint that limits one in terms of the effort that can be expended to create a given solution. Ideally, a cost-effective system must provide the most effective solution at the least cost. However, given the constraint, one has to trade effectiveness for the cost. To achieve the right balance between the two, an architect has to settle for the most effective solution that can be created within the confines of the cost that can be expended. For instance, one may desire a highly available system. There may be several choices, ranging from making a system analytically redundant that can avoid common-cause failures to having a spare that can be used to replace the part that failed. The former is several times more expensive to build compared to the latter, but the former is much more desirable for continuous operation. The right balance may be to use active redundancy, which is more expensive than having a spare but not nearly as expensive as analytic redundancy. Although this appears relatively straightforward, such decisions are not always this easy, especially in the face of uncertainty surrounding effectiveness of a solution and its projected cost and the difficulty in characterizing effectiveness and cost (National Aeronautics and Space Administration [NASA], 2007).

Because of the conflicting nature of the ASRs leading to complex trade-off situations, it is preferred only to consider the most important of these requirements, which should be only a handful. The prioritized set of half a dozen or so ASRs is often referred to as the architecture drivers.

4.2 Architectural Drivers

From the product features (functional capabilities), quality attribute scenarios, and constraints detailed in the previous chapters, a list of significant architectural drivers for the building automation system is distilled. A prioritized list of such drivers for the building automation system is shown in Table 4.1. The priorities are shown as tuples, which are described further in the text.

Architectural drivers 1 through 5 relate to the quality attribute scenarios enumerated in Table 2.6 in Chapter 2. Architectural drivers 1 and 3 also correspond to the device management and dynamic reconfiguration, 2 to the internationalization and localization, and 4 to event management and alarm management product features shown in Figure 3.4 in Chapter 3. In addition, architectural drivers 1, 3, and 5 address the product variability constraint from Table 3.4 in Chapter 3.

You should note that although quality attribute scenarios are obviously significant architectural drivers, product features

Table 4.1   Architectural Drivers for the Building Automation System Along with Their Sources and Their Priority

No.	Architectural Driver	Source	Priority
1	Support for adding new field device	• *Quality attribute scenario*: Support hardware devices from many different manufacturers • *Product feature*: Device management, dynamic reconfiguration • *Constraint*: Product variability	(H, H)
2	International language support	• *Quality attribute scenario*: Support international languages • *Product feature*: Internationalization	(M, M)
3	Nonstandard unit support	• *Quality attribute scenario*: Support conversion of nonstandard units used by different devices • *Product feature*: Device management, dynamic reconfiguration • *Constraint*: Product variability	(H, M)
4	Latency of alarm propagation	• *Quality attribute scenario*: Comply with regulations • *Product feature*: Event management, alarm management	(H, H)
5	Load conditions	• *Constraint*: Product variability	(H, H)

Note: H, high; L, low; M, medium.

and constraints can also have implied quality concerns. It is therefore important to take these into account as well. That is precisely what we have done when coming up with the list of architectural drivers in Table 4.1.

Prioritizing of the architectural drivers is done by soliciting input from both the business and the technical stakeholders.

The business stakeholders prioritize drivers based on their business value (H for high implies a system lacking that capability will not sell, M for medium implies a system lacking that capability will not be competitive, L for low implies something nice to have), whereas the technical stakeholders do so based on how difficult it would be to address a driver during the system design (H for high implies a driver that has a systemwide impact, M for medium implies a driver that has an impact on a significant part of the system, L for low implies a driver whose impact is fairly local and therefore not architecturally significant), resulting in nine different combinations in the following order of precedence: HH, HM, HL, MH, MM, ML, LH, LM, and LL.

> Architectural drivers are a prioritized set of most significant architectural requirements derived predominantly from quality attribute scenarios as well as product features and constraints that have implied quality concerns.

4.3 Patterns

Once a prioritized set of architectural drivers is established, the architecture design can begin. But, how do we proceed? The building automation system, introduced in the previous lessons, has a modifiability quality attribute requirement—a need to integrate hardware devices from many different manufacturers. More concretely, the requirement cast as a quality attribute scenario states "a field engineer should be able to integrate a new hardware device into the system at runtime and the system should continue to operate with no downtime or side effects." How do we achieve this requirement?

One approach is to look for other systems that may address a similar problem. For instance, operating systems experience

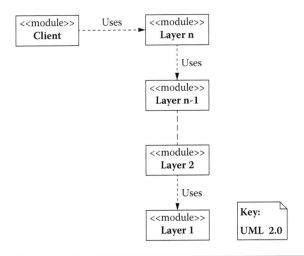

Figure 4.1 Static layered structure of an operating system.

a high degree of variability in dealing with the underlying hardware devices on the machine that run them. How do you think this variability is handled?

Operating systems use a hierarchical layered structure in which each layer presents an abstract (virtual) machine to the layer above as shown in Figure 4.1. In this figure, a client of the operating system interacts with a layer at the highest level of abstraction, and the lower-level layers manage the variability in the underlying hardware. Higher-level layers are allowed to use the services of an adjacent lower-level layer but not the other way around.

Figure 4.2 shows a typical scenario in which a client makes a request to view a file to the operating system, which in turn asks the file manager to read the file from the disk. The file manager uses the appropriate disk driver to read the disk.

In this scenario, the disk driver directly interfaces with a physical disk, thereby abstracting away from the file manager the details of the protocol to follow when communicating with the physical disk. The file manager uses an appropriate disk driver to communicate with a physical disk, thereby abstracting away from the operating system the details of how requests are routed to and responses received from (possibly) a multitude of disk drives.

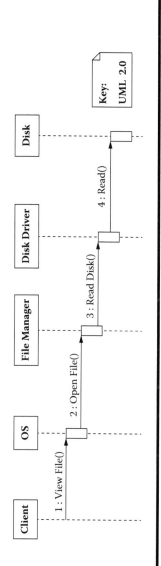

Figure 4.2 Dynamic view showing interaction among the different elements of an operating system.

Insofar as the operating system is concerned, there is a single (virtual) disk. The operating system abstracts away from the client how the underlying hardware disks are being managed.

The modifiability problem in the building automation system is similar to the scenario presented in the context of an operating system. Therefore, the provided solution can be adapted to resolving the modifiability problem as well. The example demonstrates that there can be a recurring *problem* that arises in a given *context* that can be resolved using a well-proven *solution* to the problem. That is, in the given example, the problem in the operating system domain also recurs in the building automation domain and is, therefore, resolved using a well-proven hierarchical layering structure or layers pattern.

A pattern therefore presents a well-proven *solution* to a recurring design *problem* that arises in specific design *contexts*. The structures designed by architects of software systems are often based on one or more patterns. Although there are numerous sources of architectural patterns (for software-intensive systems, see multivolume series on pattern-oriented software architecture: Buschmann et al., 1996, 2007a, 2007b; Schmidt et al., 2000; Kircher and Jain, 2004), the following sections show some classical architectural patterns used in software-intensive systems based on a classification suggested by Avgeriou and Zdun (2005). This classification is based on the predominant perspective or view an architect has of the system under design. A number of examples used for illustrating these patterns are taken from Duell et al. (1997).

> A pattern is a proven *solution* to a recurring design *problem* that arises in specific design *contexts*.

4.3.1 Layered View

In a layered view, a system is seen as a complex heterogeneous entity that needs to be decomposed into its interacting parts.

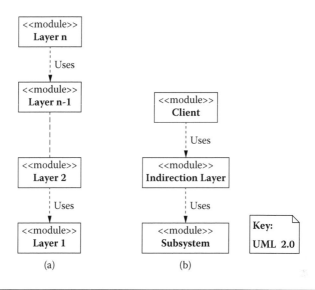

Figure 4.3 **Architectural patterns based on a layered view: (a) layers and (b) indirection layer.**

Patterns under this view deal with how these parts interact with each other while remaining as decoupled from each other as possible. Figure 4.3 shows patterns based on this perspective.

The layer pattern in Figure 4.3a decomposes a system into multiple layers of abstraction in which each layer provides a set of services to the layer above and uses the services of the layer below. A higher layer is only allowed to use an adjacent layer immediately below; bypassing a layer is not permitted. This pattern promotes modifiability, portability, and reusability in a system. Modifiability is improved because changes can be localized to a given layer, and the layer can further prevent these changes from propagating to other parts of a system by providing its services to higher layers via a standard stable interface. Portability is improved because, to deploy the system on multiple different platforms, only the platform-specific layers need to be modified. If layers represent different levels of abstraction, a subset of lower-level layers could serve as a foundational platform and be reused for developing a variable set of higher-level layers or vice versa.

The indirection layer pattern shown in Figure 4.3b introduces a layer in a system to hide a component or subsystem providing access to its services indirectly to a client of that component or subsystem. This may be necessary when a client needs to communicate with a legacy subsystem (perhaps written in a different language and running on a different platform) without becoming coupled to its implementation details.

Layering is widely used in operating systems, communication systems, and information assurance. An interesting use of layering in operating systems is to separate the most essential functions from nonessential ones. The essential functions are implemented in a layer called the kernel, and the nonessential functions are implemented in higher-level layers that use the kernel. The advantage of this separation is that the kernel ends up being fairly small and can be designed to be highly reliable. The nonessential functions outside the kernel run with only user privileges, thus preventing undesirable effects when they fail. The system is also more secure because user privileges prevent nonessential functions from accessing more privileged operations/resources that could potentially compromise the system.

Communication systems use the Open System Interconnection (OSI) model (International Organization for Standardization/International Electrotechnical Commission [ISO/IEC] 7498-1), which is a seven-layer abstraction model that standardizes how systems communicate with each other. The lowest-level *physical layer* is concerned with the specification of the physical transmission medium (cable, hubs, repeaters, adapters, etc.). The *data link layer* uses the physical layer for transmitting data between two communicating points on a network. The *network layer* uses the data link, breaking the data into variable-length sequences called datagrams and using multiple routes to deliver them to their destination. The *transport layer* utilizes the network layer, chunking data into packets and ensuring reliable delivery of these packets to their destination. The *session layer* controls the connections between two communicating points on a network. Finally, the *presentation layer* handles any differences in the syntax and semantics of data that the network delivers and what an end-user application (the *application layer*) at the destination accepts.

Defense in depth, conceived by the National Security Agency (NSA), is an information assurance pattern that uses multiple layers of security in a system such that a breach of security in any one layer can be detected and prevented by subsequent layers from cascading to other (more critical) parts of the system.

4.3.2 Data Flow View

In a data flow view, a system is seen as one that transforms streams of data. Patterns under this view describe how streams of data can be successively processed or transformed by components that are independent of one another. Figure 4.4 shows patterns based on this perspective.

Figure 4.4a shows the batch sequential pattern by which a system performs a task using a series of components with each component representing some step in the sequence of the whole task. During each step in the sequence, a batch of data input for the task is processed and sent as a whole to the next step. Figure 4.4b shows the pipes-and-filters pattern, which is similar to batch sequential except a processing component is called a filter and adjacent filters communicate with each other via connectors called pipes. Filters and pipes can be flexibly composed as adjacent filters do not have to know about each other. Filters process data incrementally, and pipes act as data buffers, creating a potential for filters to work in parallel; a filter can start placing its output incrementally on a pipe, and a filter waiting for this data can start processing as soon as the first increment arrives.

Financial applications use a wide variety of data formats, including Interactive Financial Exchange (IFX), Open Financial Exchange (PFX), and Electronic Data Interchange (EDI). When integrating these applications, the data in one format may

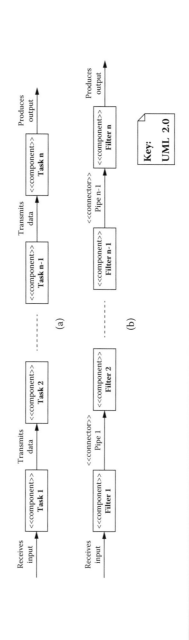

Figure 4.4 Architectural patterns based on a data flow view: (a) batch sequential and (b) pipes and filters.

need to be transformed into another format. Filters can be designed to take one of the data formats as inputs and transform it into the desired format. These filters can then be flexibly combined to achieve the desired data flow.

Water treatment plants illustrate the use of pipes and filters; water pipes play a role similar to data pipes, and various stages of treating water (removing sediment, adding chlorine, adding fluoride, etc.) correspond to filters. These filters can be changed or combined in any order, and new filters can be added. For instance, a homeowner may feed the water to a refrigerator, which may filter chemicals and cool the water for drinking.

4.3.3 Data-Centered View

In a data-centered view, a system is seen as a persistent shared data store accessed and modified by a number of components. Patterns under this view describe how a data store is accessed by multiple components. Figure 4.5 shows patterns based on this perspective.

Figure 4.5a shows the shared repository pattern in which one component of the system acts as a central shared data repository accessed by other independent components. A shared repository must handle issues associated with resource contention. A shared repository may be a passive data store or it may be active, maintaining a registry of clients and notifying

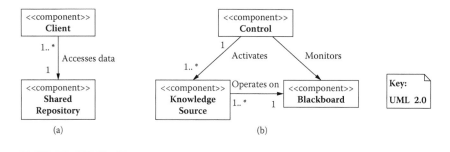

Figure 4.5   Architectural patterns based on a data-centered view: (a) shared repository and (b) blackboard.

them of specific events (such as data created, updated, or deleted) in the repository when they happen.

Figure 4.5b shows a blackboard, which breaks a complex problem for which no deterministic solution exists into smaller subproblems that can be deterministically solved by expert knowledge sources. The results of these knowledge sources are used in a heuristic computation to incrementally arrive at a solution for the complex problem. A blackboard serves as a shared repository for these knowledge sources, which use it to post results of their computation and also periodically access it to see if new inputs are presented for further processing. A control component monitors and coordinates access to the blackboard by these knowledge sources.

The blackboard pattern has been used in diverse domains, such as image and speech recognition, cryptanalysis and surveillance. For a speech recognition system, the input is speech recorded in waveform, and the output is English (or any other language) sentences. This transformation involves linguistic, acoustic, phonetic, and statistical expertise that can successfully convert acoustic waveforms into phonemes, syllables, words, phrases, and finally sentences.

Cryptanalysis leverages multiple sources, such as codebooks, translators, and traffic analysis for decrypting a coded message. The input comes in as encrypted message that is incrementally transformed by these sources into a decrypted message in English or any other desired language.

The military uses input from multiple sources, such as radar, infrared detectors, global positioning systems, communication, and electronic intelligence, for their surveillance systems. A surveillance system collects data from these diverse sources, fusing together and analyzing it for assessing a given situation.

Criminal investigations also involve a kind of blackboard pattern in which the lead detectives may use a blackboard (literally) to post forensic evidence, ballistic reports, crime scene data, and so on. They use different experts (forensic, ballistic, etc.) to piece together these data to solve the crime.

4.3.4 Adaptation View

In an adaptation view, a system is seen as a core part that is stable and an adaptable part that changes over time or in different versions of the system. Patterns under this view deal with how the system adapts itself during its evolution. Figure 4.6 shows patterns based on this perspective.

Figure 4.6a shows the microkernel pattern in which the system is structured as internal servers and external servers. Clients access the microkernel only through external servers. The internal servers are hidden from the clients and can be adapted to realize parts that need adaptation. In a gaming system, the gaming console serves as a microkernel, and the game cartridges or cards are the internal servers. The input devices of the gaming console, such as buttons, joystick, and the like, serve as external servers through which the player (client) interacts.

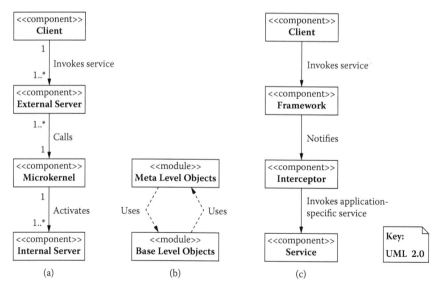

Figure 4.6 Architectural patterns based on an adaptation view: (a) microkernel, (b) reflection, and (c) interceptor.

Figure 4.6b shows the reflection pattern in which aspects of the system (structural or behavioral) that can change at any point in time are stored in meta-level objects. The base-level objects can query the meta-level objects to obtain this structural or behavioral information to execute the functionality they need. When viewing the legislative branch of the US government, the Constitution serves as a meta-level object, and the Congress is the base-level object. The Congress conducts itself according to the information contained in the Constitution, and the Constitution provides the mechanism for changing the (structure and behavior of) the Congress. Meta-level objects may also query the base-level objects and execute functionality based on their state. For instance, which rule of the Constitution applies depends on what state the Congress is in. In the event of a deadlock in the Senate, the vice president casts a vote to break the deadlock.

Figure 4.6c shows the interceptor pattern; a system is structured as a framework and a client that can extend this framework. The framework provides a client services that may change over time or with the needs of the client. A client can register with the framework, an interceptor that can intercept incoming events from the client and reroute them to the intended receiver, such as a client-specific version of a service. This way, a client can transparently update services provided by a framework and access them via the interceptors. An example of this pattern can be experienced with how the US Post Office delivers mail. A customer (client) can instruct the post office (framework) to stop the mail delivery (service) for a predetermined length of time. A postal worker (interceptor) can intercept the mail intended for the customer and hold the mail (modified service) for the instructed length of time.

4.3.5 Language Extension View

A system is viewed as a part that is native to its environment and another part that is not. Patterns under this view offer an abstraction layer to the system that translates the nonnative part into the native environment. Figure 4.7 shows patterns based on this perspective.

Figure 4.7a is the interpreter pattern where the nonnative part of a system may be in the form of scripts or a language which is an interpreter translates into a language native to a given runtime platform or execution environment. For instance, the United Nations may draft a resolution in English that the interpreters may translate into various languages native to the dignitaries or representatives from various participating nations.

Figure 4.7b is a variation of the interpreter pattern; the virtual machine in this variation defines a simple architecture on which an intermediate form of a language can be executed. The translation of an original script (in a language not native to the system) to an intermediate language is handled by a language compiler. The virtual machine itself can then translate the intermediate language into a language native to a given runtime platform or execution environment.

Figure 4.7c shows a rule-based system used for expressing logical problems in a system and consists of facts and rules and an engine that acts on them. Facts are stored in a working memory and condition-action rules in a rule base. A client may wish to know if a fact is implied by the existing rule base and already known facts or if any new facts can be derived from the existing rule base and already known facts. It submits this query (or problem) to an inference engine. The inference engine collects rules whose conditions match facts in the working memory and performs actions indicated by the rules. The actions involve adding new facts to or deleting existing facts from the working memory. The inference engine continues to do this until the problem is solved. This pattern is widely used in expert systems.

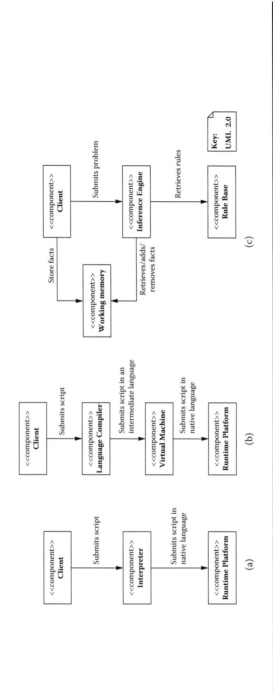

Figure 4.7 Architectural patterns based on a language extension view: (a) interpreter, (b) virtual machine, and (c) rule-based system.

4.3.6 User Interaction View

In a user interaction view, a system is seen as a part that represents a user interface and a part that represents system logic associated with that interface. Patterns under this view deal with decoupling the user interface from the system logic. Figure 4.8 shows patterns based on this perspective.

Figure 4.8a shows the model-view-controller pattern in which the system offers multiple user interfaces. It is

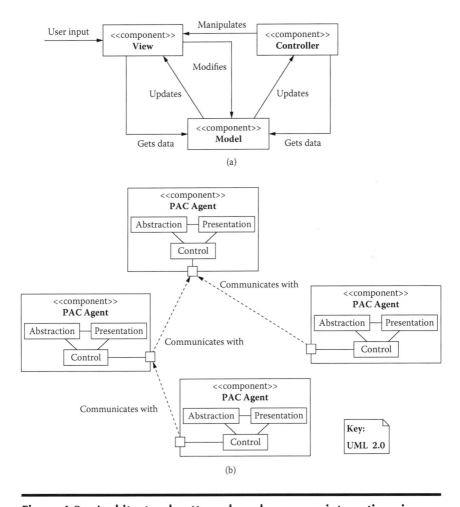

Figure 4.8   Architectural patterns based on a user interaction view: (a) model-view-controller and (b) presentation-abstraction-control.

structured as a model that encapsulates its data along with
the associated logic, one or more views that display por-
tions of this data, and a controller associated with each view
that translates any request from the view to a request for the
model. View and controller together constitute a user interface.
The users interact strictly through the user interface, and the
model in turn notifies all its associated user interfaces so they
accurately reflect all of its updates. An example of this pattern can be seen in an election year. National polls keep tally
of vote counts (model) from different regions and display the
total results (view) that reflect how each party is faring in the
election. Whenever the vote tally (model) changes, the total
results (view) are immediately updated to reflect that change.

Figure 4.8b shows the presentation-abstraction-control
(PAC) pattern in which a system is structured as a hierarchy
of multiple cooperating PAC agents that offer multiple diverse
functionalities, each with its own user interface. These functionalities, however, need to communicate with each other.
The system is therefore structured like a tree of agents:
Leaf-level node agents represent the diverse functionalities,
middle-level node agents combine the functionalities of the
lower-level agents, and the top-level node agent orchestrates
the middle-level node agents to offer their collective functionality. Each agent is structured as a presentation part that represents the user interface, the abstraction part that represents its
data along with the associated logic, and the control part that
mediates between the two and the controls of all other agents.
In the election scenario, votes may be tallied by voting districts, states, and the nation as a whole. In this case, the leaf-
level agents would correspond to the voting districts that tally
the votes and display results of the local government elec-
tions. They, however, communicate with their respective states
(middle-level agents), which aggregate votes for state representatives and display the results for the state. The states in turn
communicate to the nation as a whole (top-level agent), with

votes aggregated and displayed for the national government. Thus, details can be viewed at each level.

4.3.7 Component Interaction View

In the component interaction view, a system is viewed as a collection of independent components that interact with each other in the context of the system. Patterns under this view focus on how individual components exchange messages while retaining their autonomy. Figure 4.9 shows patterns based on this perspective.

Figure 4.9a shows the explicit invocation pattern; a client component explicitly invokes the services of a server component either synchronously or asynchronously. The client must know about the location of the server, the name of the service being invoked, and its parameters. The client-server pattern is a variant of explicit invocation, by which the server needs to scale because of requests coming from multiple clients. Together, the clients and server must implement tasks such as session, security, and transaction management. As an example, customers (clients) can call in to a department store (server) to order products. They need to know the 800 number (location) to call; the product and the department from which it is to be ordered (service); and the details of the product and their shipping and billing information (parameters). The department store needs to understand how many client calls are to be handled on a given day (scalability needs) and provide enough associates to take these calls. The store must ensure the security of the transaction and guarantee the handling of the transaction (order fulfillment and shipping) or no money will be charged to the clients.

Another variant of explicit invocation is the peer-to-peer pattern in which there is no distinction between client and server components. They can serve both roles and request services of each other. For example, in peer-to-peer market-places, people can purchase goods from others offering them

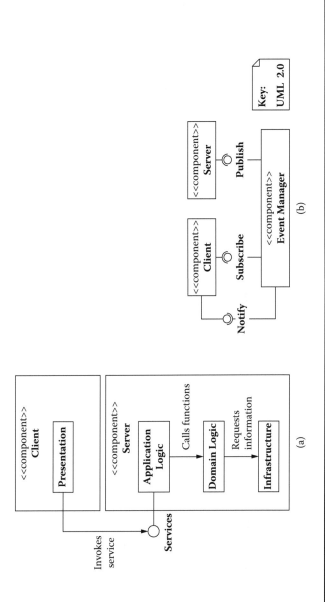

Figure 4.9 **Architectural patterns based on a component interaction view: (a) explicit invocation and (b) implicit invocation.**

for sale. Those who sell these goods can in turn purchase items offered for sale by the very people who bought from them. So, anyone can play a role of a client or a server in this scenario.

In the implicit invocation pattern, a client component implicitly invokes the services of a server component. The client and server are decoupled because the invocation happens indirectly. Figure 4.9b shows the publish-subscribe pattern, which is a variant of implicit invocation; in this variant, clients (event consumers) subscribe to specific events, and servers (event producers) publish specific events. The publishing of an event triggers a callback mechanism that notifies the subscribers or consumers of that event. Auctioning is a good example of this pattern. An auctioneer (event manager) broadcasts, from a current bidder (publisher), the bid for an item being auctioned to the other bidders (subscribers). If one of the bidders accepts this bid, then he or she becomes the current bidder (publisher) and the auctioneer (event manager) publishes his or her new bid to the other bidders (subscribers).

4.3.8 Distribution View

A system is viewed as a collection of components that are distributed among different network nodes or processes. Patterns under this view deal with interactions among distributed components. Figure 4.10 shows patterns based on this perspective.

Figure 4.10a shows a broker that separates and hides the communication functionality from the rest of the system and mediates all communication among its distributed components. For instance, when planning travel, a customer may use a travel agency for booking services such as air, hotel, and car. The customer (client) may use the travel agency system (client-side proxy) for making arrangements for these services, and the travel agency (broker) may use a reservation system (server-side proxy) for booking these services (server).

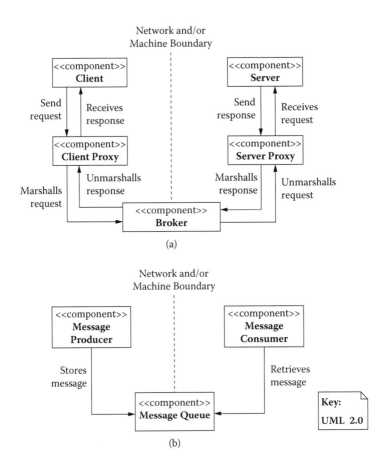

Figure 4.10 Architectural patterns based on a distribution view: (a) broker and (c) message queuing.

Figure 4.10b shows a message queuing pattern in which messages between components of a system are exchanged through message queues. The sender of a message simply deposits a message in a queue and the receivers monitor the queue, picking up the messages they are interested in when they arrive. For instance, waiters (message producers) in a restaurant place orders from their customers in a queue (message queue), and the bartenders or chefs (message consumers) can monitor these queues and pick up the orders to be filled. They in turn can play the role of message producers, marking the orders filled and placing them in a queue (message queue) that waiters (message consumers) monitor, picking up filled orders to be delivered to their customers.

> Patterns provide well-known solutions to recurring design problems that arise in a given context.

4.4 What Is a Tactic?

Patterns are composed from fundamental design decisions known as tactics. Tactics, therefore, are the building blocks of design from which architectural patterns are created. For instance, when thinking of the layer pattern introduced previously, we notice that it entails several design decisions. It uses the following:

- *Semantic coherence* to assign to each layer those responsibilities that work closely together without excessive reliance on other layers. This increases cohesion within a layer and minimizes its coupling to other layers, with the goal of localizing the impact when modifications are made to a given layer (things that work together are likely to change together).
- *Hide information* to make services of a layer available through interfaces keeping internal details of how these services are provided by the layer private with the goal of preventing changes made to a given layer from propagating to the other layers (things that are not publicly visible can change and evolve independently).
- *Limit exposure* by placing layers as intermediaries. The intermediaries can thus decouple higher-level layers from lower-level layers, limiting the exposure of some critical design decisions (limiting the exposure of security-critical functions can prevent them from becoming a target of attack).

Semantic coherence, hide information, and limit exposure are tactics that together yield the layer pattern. This example also illustrates that a given tactic is targeted; it focuses on a design concern related to a single quality attribute: Semantic coherence addresses the modifiability concern related to localizing the impact of change; information hiding addresses the modifiability concern related to preventing the ripple effect of a change; and limit exposure addresses the security concern related to resisting/preventing attacks. A pattern, on the other hand, can address multiple concerns related to multiple quality attributes at the same time; the layers pattern addresses multiple design concerns related to modifiability and security.

In the following sections, a subset of tactics based on the work of Bass et al. (2013) is described. These tactics address design concerns related to some predominant quality attributes and Bass et al. (2013) provide a more complete catalog of these tactics.

4.4.1 Tactics for Availability

The availability quality attribute is concerned with managing systems faults. If left unmanaged, system faults can cause failures that may compromise the system's ability to deliver certain services (some of which may be safety critical and could lead to loss of life) or render the system completely unusable. Availability, therefore, is concerned with managing faults such that system services continue to operate as expected, making a system more reliable, secure, and safe. It is measured in terms of the length of time for which a system continues to operate without experiencing any downtime that results from a failure, referred to as mean time between failures or MTBF, and the length of time it takes to recover should a system experience a failure, referred to as mean time to repair or MTTR. The ratio (MTBF/(MTBF + MTTR)) yields the availability percentage of a system. The system is 100% available if MTTR is 0 (it takes no time to recover or, more important, faults never become failures). A system that has availability of

99.999% experiences a downtime of a little over 5 min each year as a result of failures.

Tactics for availability are concerned with adding capabilities to a system for handling faults. Table 4.2 provides a subset of these tactics categorized by the availability design concerns they address.

For handling faults, an important design concern is the ability to *detect the occurrence of a fault*. The *monitor* tactic uses a process that continuously monitors the status of different elements of a system. The monitoring process may use an additional tactic called *ping/echo* to periodically ping an element for its status and could receive an echo back from the element indicating it was operating normally. Alternatively, the monitoring process may use a *heartbeat* tactic; it waits for a periodic status update from the monitored element, and if this update does not arrive, it assumes a fault has occurred. The periodic status update may simply be the monitoring process observing messages flowing back and forth from the element being observed and assuming a fault if the element has been inactive for some predetermined amount of time. An element

Table 4.2 Tactics for Addressing Availability Design Concerns

Design Concern	Tactic
Detect the occurrence of a fault	• Monitor
	• Ping/echo
	• Heartbeat
	• Exception
Recover from a fault	• Active redundancy
	• Passive redundancy
	• Checkpoint/rollback
	• State resynchronization
	• Voting
Prevent the occurrence of a fault	• Removal from service
	• Transactions

can also use an *exception* tactic and throw an exception to let the monitoring process know an exception has occurred.

Once a fault is detected, the monitoring process may need to repair the system so that the fault does not manifest itself as a failure. This is especially true when dealing with safety-critical systems, for which a failure could lead to a catastrophic loss. So, an important design concern for building highly available systems is to *recover from a fault*. An *active redundancy* tactic suggests introducing a redundant copy of a safety-critical element into the system. Both the safety-critical element (called the primary) and its redundant copy (called the secondary) actively process their inputs, but the system utilizes the output of the primary only. When a fault occurs, the monitoring process detects the fault, promotes the secondary to the primary, and reconfigures the system to start using the new primary. The monitoring process then initiates the recovery of the failed primary and on successful recovery reintroduces it into the system as a secondary. The *passive redundancy* tactic is similar to active redundancy except only the primary element actively processes all input while the secondary lies dormant. When the primary fails, the monitoring process must do the following:

- Activate the secondary, which brings itself into a state consistent with the primary before the primary had failed.
  - To achieve this, the primary must use a *checkpoint* tactic to periodically save its current state, and the secondary can use a *state resynchronization* tactic to read the primary's checkpoint state, bringing itself into a state consistent with that of the primary before the primary had failed.
- Promote the secondary element to be the new primary and reconfigure the system to start using the new primary.
- Run recovery for the failed primary and reintroduce it as a secondary into the system after its recovery is completed successfully.

When redundant elements are identical copies of each other, there is a distinct possibility that the cause that led to the failure of the primary may also cause the secondary to fail. To avoid such common-cause failures, functional redundancy or analytic redundancy is used. In functional redundancy, the redundant elements process the same inputs and produce the same outputs, but they do so using different processing algorithms (for instance, a set of linear equations can be solved using different mathematical techniques, such as substitution or elimination). In analytic redundancy, the redundant elements process different inputs using different processing algorithms and produce different outputs (for instance, an aircraft can determine its altitude using a barometer or radar). When using such diverse inputs, processing algorithms, and outputs, a more sophisticated tactic called *voting* is used to determine if a component has failed. An element called a voter monitors the output of all redundant components and uses a sophisticated (domain-specific) algorithm to identify and overcome failures.

When a fault is detected, action must also be taken such that the fault does not propagate to other parts of the system. Therefore, another availability design concern is to *prevent the occurrence of a fault*. A tactic such as *removal from service* simply removes a failed element from service to prevent any cascading effect of its failure. Another tactic called *transactions* bundles together several tasks into a unit of work called a transaction and ensures that either the entire transaction completes or any effects of partial completion (which may leave a system in an inconsistent state) are completely undone.

4.4.2 Tactics for Interoperability

The interoperability quality attribute concerns itself with the ability of two or more systems to work together to achieve some given purpose. Often, these systems may not be under a centralized control and may have been developed without

Table 4.3 Tactics for Addressing Interoperability Design Concerns

Design Concern	Tactic
Locate a service	• Discover service
Manage interfaces to a service	• Orchestrate • Tailor service

envisioning any interoperability scenarios. Under these circumstances, mechanisms have to be put in place such that a system can discover the services it needs from another system and can use these services to have a meaningful exchange. Tactics for interoperability are designed to make this possible. Table 4.3 provides a subset of these tactics categorized by the interoperability design concerns they address.

Before a system can use one or more services offered by another system, it must know about their location. This *locate-a-service* interoperability design concern is addressed via the *discover service* tactic, which suggests putting in place a directory that systems can use to register the services they provide and that other systems can query to discover the systems that provide the needed services. Once the services have been discovered, they have to be invoked via some interface so the work needed is done. This *manage-interfaces-to-a-service* design concern can be addressed using two different tactics. The *orchestrate* tactic suggests having a mechanism to compose the invocation of a diverse set of services (perhaps across multiple different systems) into a workflow, and the *tailor service* tactic suggests having an ability to adapt the interface of the invoked services to conform to a given need.

Having a standardized location, orchestration and adaption mechanisms create location and platform transparency that can significantly reduce the effort required to bring together a diverse set of systems possibly distributed across machine and network boundaries and perhaps running on multiple different platforms.

4.4.3 Tactics for Modifiability

Modifiability is a quality attribute concerned with the ease of making a change to a system. Systems that are easier to modify are more maintainable, easy to extend, and easy to port to multiple different platforms and offer greater opportunity for reuse. Improved modifiability leads to a more modular system, increasing its adaptive capacity. A system's adaptive capacity is important for its ability to evolve gracefully and has immense economic value (Parnas, 1972; Baldwin and Clark, 2000; Sullivan et al., 2001).

Tactics for modifiability deal with minimizing the effort required for making changes to a given system. Table 4.4 provides a subset of such tactics categorized by the modifiability design concerns they address.

The foremost modifiability design concern is *limiting the impact of a change* made to a system; that is, designers strive to localize a change to one and only one place in a given system so that its impact is limited. The tactic to *maintain*

Table 4.4   Tactics for Addressing Modifiability Design Concerns

Design Concern	Tactic
Limit the impact of a change	• Maintain semantic coherence • Abstract common services • Generalize the module • Anticipate expected changes
Limit the ripple effect of a change	• Restrict dependencies • Use an intermediary • Hide information • Maintain existing interface
Defer the binding time of a change	• Polymorphism • Runtime registration • Configuration files • Adhere to defined protocols

semantic coherence suggests that system elements that are used together should belong together because elements that are used together are likely to change together. Therefore, bundling such elements into a single module not only makes the system semantically coherent (easier to understand) but also localizes any changes made to these elements, making such a change more manageable. In the same vein, the tactic to *abstract common services* suggests factoring into common services the aspects that are similar across multiple elements of a system. By abstracting these common services into their own modules, changes to these services are again localized to these modules rather than distributed across multiple elements. The tactic *generalize module* advocates increasing the ability of a module to handle a much broader set of inputs, thus eliminating the need for having several specialized modules to handle subsets of these inputs. The tactic to *anticipate expected changes* suggests we should determine in advance the changes that are likely to be made to a given system in the foreseeable future so that we can prepare the system for such changes. Stated another way, we do not have to make each and every thing within the system more modifiable, only those that are likely to change.

When an element within a system is changed, it is likely that other elements that depend on the one being changed are also affected. So, once anticipated changes to a system have been localized, the next design concern is *limiting the ripple effect of a change*. One tactic to achieve this is to *restrict dependencies*, which suggests designing a system in a way such that the number of elements depending on an element likely to change is minimized. Another tactic is to *use an intermediary*, which decouples an element likely to change from the rest of the system, and all access to this element is channeled through an intermediary. Therefore, in the likelihood that a change to the element does ripple, its effect would be muted by the intermediary. Because some elements must still depend on an element likely to change (a completely decoupled element would sit by itself and would serve no useful purpose for a system), the *hide information* tactic suggests that the changeable element must hide as much internal detail as possible from the external elements with which it interacts. These external elements should interact with the changeable element through interfaces that only make public the services that these elements need. Furthermore, the tactic to *maintain existing interface* suggests that these public interfaces should be stable so that external elements that depend on them are not affected.

Once the changes have been localized and their ripple effect contained, one has to worry about who will make the change and when. It is likely that a change is made by engineers during the development cycle of a system, or it is made by the installers at the time the system is being deployed, or the change is made by the technicians while the system is operational. Therefore, the next design concern is *deferring the binding time of a change*, that is, designing the architecture of a system so that it is capable of delaying when a change is made to the system. To make this possible, the *polymorphism* tactic can be used to make an element closed to modification but open to extension; that is, the element itself need not change (hence be closed to modification), but its capabilities can be enhanced through one or more extension points it provides (hence be open to extension). If these extensions need to be bootstrapped while a system is operation, then a combination tactic of *runtime registration* and *configuration files* can be used. When a new capability is added while the system is in operation, the system can detect this change and read an accompanying file that contains configuration parameters indicating to the system how to bootstrap this additional capability. The strategy to *adhere to defined protocols* suggests using a standard protocol, in which case the system can simply use a standard protocol for interacting with an extension when it detects this additional capability.

4.4.4 Tactics for Performance

The performance quality attribute is concerned with decreasing latency and increasing throughput. Latency is decreased if the system is responsive or can respond to a request rapidly. Throughput is increased if a system can process a high volume of requests without any perceptible decrease in its responsiveness. Addressing latency and throughput is important because it has an impact on how quickly end users can make a system process their individual requests. Responsive systems are not only scalable (can handle a high volume of requests) but also improve a system's availability (by not choking in the face of a high volume of requests).

Tactics for performance deal with improving the responsiveness of a system. Table 4.5 provides a subset of these tactics categorized by the performance design concerns they address.

The first design concern when improving performance of a system is to *manage the demand for its resources.* For software-intensive systems, this means demand for any shared resource such as central processing units (CPUs), memory,

Table 4.5   Tactics for Addressing Performance Design Concerns

Design Concern	Tactic
Manage demand for a resource	• Increase resource efficiency • Reduce overhead • Prioritize events • Bound execution times • Manage sampling rate • Limit event response
Manage a resource	• Increase resources • Introduce concurrency • Maintain multiple copies • Schedule resources

network, or data. The higher the demand for these resources, the more contention it will generate, affecting the overall responsiveness of a system. The tactic to *increase resource effi-ciency* suggests looking at the ways algorithms use a resource to process an event or input and optimizing them to reduce their processing time. Further, the tactic to *reduce overhead* suggests that if this processing is accomplished by multiple processing elements distributed across process, machine, or network boundaries, then colocating them within the same process or a machine would significantly reduce the process-ing overhead created by interelement communication. The *prioritize events* tactic recommends giving higher priority to highly important events or inputs and processing them ahead of low-priority events (which may be ignored or may wait an arbitrarily long time before being processed). The *bound exe-cution time* tactic advocates, when possible, fixing the amount of time allocated to processing an event. This could compro-mise how accurate the processing result is and therefore can only be done when a less-than-accurate output is acceptable. If a system is becoming overwhelmed by the number of events it needs to process because it is sampling these events from its environment at a certain rate, then the tactic to *manage sam-pling rate* suggests the system can reduce this sampling rate and the *limit event response* tactic advocates setting a maxi-mum event-processing rate after which events may be dropped or some policy crafted to handle the overflow.

If system performance continues to be an issue after having utilized all possible tactics to manage demand for a resource, then the next design concern to address would be to *manage a resource* itself. If cost is not a concern, the tactic to *increase resources* suggests simply adding more of a given resource, and the tactic to *introduce concurrency* recommends utilizing the capacity of a given resource by processing requests in paral-lel. The *maintain multiple copies* tactic advocates that conten-tion for a shared resource can be decreased by having multiple copies of that resource. For instance, client-server architectures replicate services clients need across a cluster of servers and use a load balancer to distribute the requests from the clients across this cluster. The *schedule resources tactic* recommends using a scheduling policy (such as first in/first out, deadline monotonic, rate monotonic, round-robin, etc.) that assigns a priority to requests for a resource and dispatches these requests in priority order when the resource becomes available.

4.4.5 Tactics for Security

The security quality attribute concerns itself with preventing unauthorized access to a resource (for instance, information or a service). It explores ways in which a system can guarantee confidentiality (a resource can be accessed only by authorized users), integrity (a resource will not be altered by unauthorized users), and availability (continued access by authorized users) of resources.

Tactics for security are designed for making systems easy to secure. Table 4.6 provides a subset of these tactics categorized by the security design concerns they address.

Table 4.6   Tactics for Addressing Security Design Concerns

Design Concern	Tactic
Resist attacks	• Authenticate users • Authorize users • Encrypt data • Limit access • Limit exposure
Detect attacks	• Detect intrusion • Detect service denial • Verify message integrity • Detect message delay
Recover from attacks	• Maintain audit trails • Restore using availability tactics

Resisting attacks is an important security design concern, and tactics that address this concern prevent access to resources from unauthorized users. The *authenticate users* tactic is used for identifying users and guaranteeing that they are who they claim to be. The *authorize users* tactic ensures that authenticated users have access to only those resources that they are authorized to access. The *encrypt data* tactic recommends keeping information in an encrypted form such that only authorized users who hold the key to decrypt such information can manipulate it. This tactic is useful when an authorized user needs to transmit confidential information to another party over a network. The *limit access* tactic advocates limiting access to privileged resources by putting in place protection mechanisms (such as Demilitarized Zone and firewalls that protect an organization's intranet). The *limit exposure* tactic suggests limiting the number of access points to resources, thereby limiting opportunities for unauthorized access.

Despite putting in place tactics for resisting attacks, unauthorized users try to gain access to privileged resources by exploiting some system vulnerability. Therefore, another important security design concern is to *detect attacks* from unauthorized users. The *detect intrusion* tactic suggests monitoring traffic coming into a system against known patterns of attacks. The *detect service denial* is a tactic for detecting a special type of attack in which the system becomes overwhelmed with a flood of unauthorized access attempts and cannot do anything useful (other than just thwarting such attempts). The tactic to *detect message delay* detects a man-in-the-middle attack by which an unauthorized user may intercept a message and alter it (therefore introducing a delay) before it arrives at the system. The tactic to *verify message integrity* uses techniques (such as hash values and checksums) to ensure that a message (or any other privileged resource) has not been altered.

If unauthorized users are successful in compromising a system, then steps must be taken to *recover from the attacks*.

Maintain audit trail is a tactic that keeps a trail of all activity on the system and can be used not only to identify and prosecute attackers but also to improve the security of the system in the future. If the system becomes compromised, then it can be restored by performing *recovery using the availability* tactics.

4.4.6 Tactics for Testability

Testability is a quality of a system that makes it easier to test. The ease of testing hinges on how much effort is required to understand or discover the cause for faulty system behavior. Tactics for testability are designed to reduce this effort. Table 4.7 provides a summary of these tactics.

To minimize the effort required to understand or discover the cause for faulty system behavior, the foremost design concern is the ability to *control and observe system state*. The *specialized interface* tactic recommends creating specialized testing interfaces that allow gaining access to the internal state of a system. The *record/playback* tactic suggests putting in place a mechanism that records any information that crosses a system interface. This information can then be used as an input for playback or exercising the same interface and therefore reproducing the (possibly faulty) behavior of the system. The *built-in monitor* tactic advocates that, to improve the testability of an element within a system, the element can have a built-in monitor that maintains and makes available its internal state, thus providing increased visibility of its internal operations.

Table 4.7   Tactics for Testability

Design Concern	Tactic
Control and observe system state	• Specialized interface • Record/playback • Built-in monitor

4.4.7 Tactics for Usability

Usability is the quality of the system concerned with its ease of use. Improved usability makes systems easier to learn, navigate, and work with; systems can guide their users through the workflows, seeking their input at each step, providing feedback, and preventing them from making any errors. Usability tactics therefore are focused on reducing the effort needed in gaining proficiency in the use of such systems. Table 4.8 provides a summary of such tactics categorized by the usability design concern they address.

To make a system easier to use, one of the usability design concerns is *supporting user initiative* or allowing the user to initiate certain actions. For instance, after launching a particular task, users may change their mind (for various reasons, such as the task is taking too long to complete or the user simply made a mistake) and may want to abort the task. The *cancel* tactic suggests putting in place a mechanism within the system that allows a user to terminate a task. Terminating a task may not be as simple as just ending the task midstream; all the work the task performed before aborting may have caused side effects that could leave a system in an inconsistent state. The *undo* tactic recommends putting in place a mechanism that undoes these side effects. The *pause/resume* tactic advocates having an ability to pause a long-running task

Table 4.8   Tactics for Addressing Usability Design Concerns

Design Concern	Tactic
Support user initiative	• Cancel • Undo • Pause/resume
Support system initiative	• Maintain task model • Maintain user model • Maintain system model

midstream and resume it at a later time (so that a user is not compelled to cancel it simply because it is taking too long).

After launching a task, it would also be nice for users to track its progress and receive guidance in successfully completing the task. Therefore, the next usability design concern is *support system initiative* or support for the ability of a system to provide periodic feedback to the user. The *maintain system model* tactic advocates creating a model that allows a system to determine its expected behavior and gives it the ability to predict the completion time (by displaying a progress bar, for instance) of a currently running task. The *maintain task model* tactic recommends having a model of the task itself so the system has a context of what a user is attempting to accomplish and can provide useful feedback by identifying any errors (by highlighting incorrect inputs or correcting them when possible) and can help navigating through the workflow (by going to the next logical step in the workflow where user input is needed). The *maintain model of the user* tactic suggests keeping a model of a user so that the system can provide guidance based on whether a user is a novice or an expert, for instance.

> Patterns are composed of fundamental building blocks of design called tactics.

4.5 Summary

This chapter began to explore the shaded portion of ISO/IEC 42010 conceptual framework for system and software architecture shown in Figure 4.11.

In particular, we discussed how the most significant architectural requirements that drive the design of a system architecture are distilled from its quality attribute scenarios and functional responsibilities and constraints with implied quality concerns.

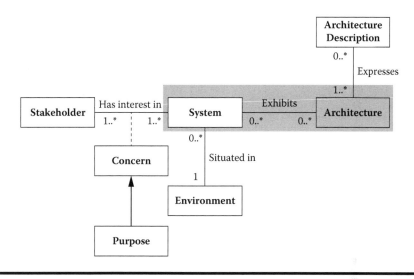

Figure 4.11   Conceptual framework for system and software architecture.

Their conflicting nature limits us to consider only a handful of these ASRs, which we refer to as architectural drivers.

Rather than designing a system architecture from scratch, architects use design patterns and tactics to satisfy the design concerns inherent in the architectural drivers. Design patterns are solutions to recurring design problems in a given context, and tactics are more fundamental design decisions that serve as building blocks for these patterns. This chapter explored a subset of these patterns and tactics.

4.6 Questions

1. Can you think of additional constraints (related to organization, technology, or product) that may have a strong influence on the architecture of the building automation system?
2. Consider the publish-subscribe pattern. What qualities might this pattern try to address? What tactics make up this pattern?

3. Although the patterns and tactics listed in this chapter relate predominantly to software-intensive systems, can you find examples of other kinds of systems in which these have been or can be applied?

4. Consider a warehouse management (WM) system that needs to communicate with an enterprise resource planning (ERP) system to control and optimize the material flow within a warehouse. The two systems are deployed across a computer network to meet their operational requirements. Consequently, there could be a machine boundary between the two.

   a. How can we shield the objects of the system from dealing with networking issues directly while supporting location-independent interaction between them? Consider the messaging or broker pattern described in the text when coming up with a solution. Clearly explain the trade-offs when making your choice.

   b. Present a sketch of your design.

   c. Identify the tactics used by the pattern of your choice.

References

P. Avgeriou and U. Zdun, Architectural patterns revisited—A pattern language. In *Proceedings of 10th European Conference on Pattern Languages of Programs (EuroPlop 2005)*, Irsee, Germany, 2005, pp. 1–39.

C. Baldwin and K. Clark, *Design Rules: The Power of Modularity,* Volume 1. Cambridge, MA: MIT Press, 2000.

L. Bass, P. Clements, and R. Kazman, *Software Architecture in Practice,* third edition. Boston: Addison-Wesley, 2013.

F. Buschmann, K. Henney, and D. Schmidt,. *Pattern-Oriented Software Architecture: A Pattern Language for Distributed Computing*, Volume 4. New York: Wiley, 2007a.

F. Buschmann, K. Henney, and D. Schmidt, *Pattern Oriented Software Architecture: On Patterns and Pattern Languages*, Volume 5. New York: Wiley, 2007b.

F. Buschmann, R. Meunier, H. Rohnert, P. Sommerlad, and M. Stal, *Pattern-Oriented Software Architecture: A System of Patterns*, Volume 1. New York: Wiley, 1996.

M. Duell, J. Goodsen, and L. Rising, Non-software examples of software design patterns. In *Addendum to the 1997 ACM SIGPLAN Conference on Object-Oriented Programming, Systems, Languages, and Applications (Addendum)* (OOPSLA '97). New York: ACM, 1997, pp. 120–124.

ISO/IEC/IEEE Systems and software engineering—Architecture description, ISO/IEC/IEEE 42010:2011(E) (Revision of ISO/IEC 42010:2007 and IEEE Std 1471-2000), pp. 1–46, 2011.

ISO/IEC JTC1, Information Technology—Open System Interconnection—OSI Reference Model: Part 1—Basic Reference Model, ISO/IEC 7498-1, 1994.

M. Kircher and P. Jain, *Pattern-Oriented Software Architecture: Patterns for Resource Management*, Volume 4. New York: Wiley, 2004.

National Aeronautics and Space Administration (NASA), *Systems Engineering Handbook* (NASA/SP-2007-6105 Rev1). Washington, DC: NASA Headquarters, December 2007.

D. Parnas, On the criteria to be used in decomposing systems into modules. *Commun. ACM* 15, 12 (December 1972), 1053–1058, 1972.

D. Schmidt, M. Stal, H. Rohnert, and F. Buschmann, *Pattern-Oriented Software Architecture: Patterns for Concurrent and Networked Objects,* second edition, Volume 2. New York: Wiley, 2000.

K. Sullivan, W. Griswold, Y. Cai, and B. Hallen, The structure and value of modularity in software design. In *Proceedings of the 8th European Software Engineering Conference Held Jointly with 9th ACM SIGSOFT International Symposium on Foundations of Software Engineering (ESEC/FSE-9)*. New York: ACM, pp. 99–108.

Chapter 5

Creating the Architecture

5.1 Introduction

Chapter 4 discussed patterns and tactics that can be used for creating an architecture of a system. We can take the list of architectural drivers for a given system and, for each driver, consider patterns or tactics that might address design concerns inherent in that driver. When the first driver on the list is considered, it will yield a structure (based on patterns or tactics selected) that will satisfy its design concerns. We can then move to the next driver on the list and examine if the structure resulting from satisfying the first driver also satisfies the design concerns of this driver. If it does, we can move on to the next driver, but if it does not, we must examine how we refine this structure (using additional patterns or tactics) to satisfy these concerns. We continue to repeat this process until the structure has been sufficiently refined to yield a final architecture that addresses all design concerns associated with all the drivers on the list. It should be noted that design concerns are seldom completely addressed. Conflicting design concerns have to be traded against each other and reasonable compromise has to be reached. Hence, creating an architecture is about finding the middle ground by trading off competing and conflicting objectives.

The attribute-driven design (ADD) method (Bass et al., 2013; Woods, 2007), developed at the Software Engineering Institute (SEI), is one approach that follows a recursive refinement process similar to the one outlined. ADD requires as input (a) a prioritized set of architectural drivers and (b) the functional responsibilities of the system. In general, ADD starts with the highest-priority driver and picks the element of the system that is most relevant to address the design concerns of the driver. In a green field project, that element would be the system itself (at the beginning of such a project, the system is a black box that has not gone through any refinement) and is also assumed to be the one that fulfills all of the functional responsibilities given as input. This element is then refined or decomposed using appropriate patterns or tactics to satisfy the given architectural driver, and its functional responsibilities are reallocated among the child elements resulting from this decomposition. ADD then moves to the next high-priority architectural driver following the same refinement process all over again until all drivers have been considered and a final architecture reached that reflects a reasonable balance for all design concerns.

It should be noted that the act of decomposing elements to create new elements gives rise to additional responsibilities. These responsibilities would stem from the newly created elements addressing design concerns associated with the architectural driver they satisfy.

The following sections illustrate the application of the ADD method to the building automation system (Sangwan et al., 2008) introduced in the previous chapters.

5.2 Architecture of the Building Automation System

For ease of cross-referencing, the architectural drivers and functional responsibilities of the building automation system are reproduced from the previous chapters in Tables 5.1 and 5.2. Recall that architectural drivers were derived from quality attribute scenarios and functional responsibilities and constraints with implied quality concerns (see Section 4.2). The functional responsibilities were derived from the system's use cases (see Section 3.3).

We start with a monolithic system as shown in Figure 5.1. We use standard Unified Modeling Language (UML) 2.0 notation and two views of the architecture. Figure 5.1a is a static module decomposition view that shows a decomposition hierarchy of modules (elements that represent coherent units of implementation) that make up the system. Figure 5.2b is a dynamic component and connector view that shows components (independently deployable and configurable elements) of the system and the connectors (communication links between components) that allow the components to interact with each other and exchange information. There may not necessarily be one-to-one mapping between modules and components;

Table 5.1 Architectural Drivers and Functional Responsibilities of the Building Automation System

	Architectural Driver	
No.	*Description*	*Priority*
1	Support for adding new field device	(H, H)
2	International language support	(H, M)
3	Nonstandard unit support	(H, M)
4	Latency of alarm propagation	(H, H)
5	Load conditions	(H, H)

Note: H, high; M, medium.

Table 5.2   Functional Responsibilities of the Building Automation System

	Functional Responsibilities
1	Send commands to a field device
2	Receive events from a field device
3	Perform semantic translation for device data
4	Configure a field device
5	Route data to a field device
6	Evaluate and execute automation rules
7	Send automation commands
8	Generate alarm notifications
9	Display device data
10	Capture/relay user commands
11	Display alarm notifications
12	Edit/create automation rules
13	Retrieve data from a field device
14	Store field device configuration
15	Propagate change-of-value notifications
16	Authenticate and authorize users
17	Persist rules, user preferences, alarms

several modules may be deployed together as a component, or a module may be deployed across several components. When this is the case, a mapping between the two is provided in the narrative; otherwise, one-to-one mapping should be assumed.

Figure 5.1a shows BAS (Building Automation System) as the only module that makes up the building automation system, and Figure 5.1b shows BAS as the only component that interfaces with the field devices it manages and reads the initial configurations of the devices from a set of configuration files. The monolith, as the only system component, is responsible

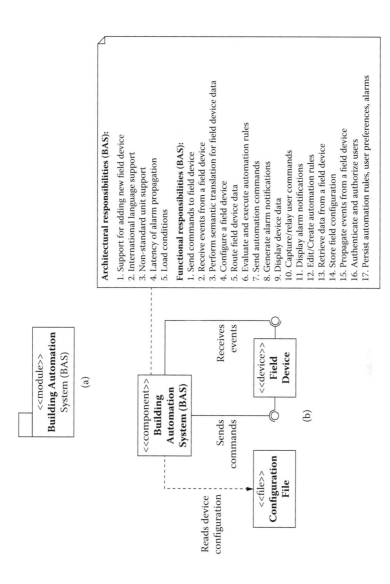

Architectural responsibilities (BAS):

1. Support for adding new field device
2. International language support
3. Non-standard unit support
4. Latency of alarm propagation
5. Load conditions

Functional responsibilities (BAS):

1. Send commands to field device
2. Receive events from a field device
3. Perform semantic translation for field device data
4. Configure a field device
5. Route field device data
6. Evaluate and execute automation rules
7. Send automation commands
8. Generate alarm notifications
9. Display device data
10. Capture/relay user commands
11. Display alarm notifications
12. Edit/Create automation rules
13. Retrieve data from a field device
14. Store field configuration
15. Propagate events from a field device
16. Authenticate and authorize users
17. Persist automation rules, user preferences, alarms

<<module>>
Building Automation
System (BAS)

(a)

<<component>>
**Building
Automation
System (BAS)**

Receives
events

<<device>>
**Field
Device**

(b)

Sends
commands

Reads device
configuration

<<file>>
**Configuration
File**

Figure 5.1 (a) Module decomposition view and (b) component-and-connector view of the monolithic building automation system.

for all functionality (Table 5.2) to be implemented along with satisfying all the architectural drivers (Table 5.1). We show this allocation of responsibilities only in the component-and-connector view in Figure 5.1b (this choice is arbitrary, and the module decomposition view in Figure 5.1a could easily have been used). This convention is followed throughout this example.

Obviously, the monolithic structure cannot satisfy most of these architectural drivers, and we need to decompose it further until all of these drivers are satisfied. We begin with one of our highest-priority architectural drivers.

5.2.1 Support for Adding New Field Devices

Recall that this driver to provide support for adding new field devices is concerned with a field engineer being able to integrate a new field device into the building automation system while the system is operational. The system should continue to operate and experience no downtime or side effects. Having this dynamic reconfiguration and device management feature supports product variability so devices of varying capabilities from many different manufacturers can be integrated into the system.

As seen in previous discussions, these types of requirements relate to the modifiability quality attribute, the ease with which a new hardware device can be integrated into a system while it is in operation. To satisfy these requirements, we can apply the modifiability tactics to limit the impact of change and minimize the number of dependencies on the part of the system responsible for integrating new hardware devices. There are really three design concerns to be addressed:

■ *Limit the impact of a change*: Our interest is to minimize the number of changes that need to be made when adding a new field device.
■ *Limit the ripple effect of a change*: We would also like to minimize the number of dependencies on those elements

that need to change as a result of adding a new field device; this would minimize the change from propagating to other parts of the system.

■ *Defer the binding time of a change*: A new field device should be deployed when the system is in operation, and the field engineers or nonprogrammers should be able to manage such deployment.

We address these concerns by decomposing BAS into adapters for field devices, a tactic of *anticipation of expected changes*; this limits the impact of change to just the adapter. We use two additional architectural tactics to limit the ripple effect of change. First, we specify a standard interface to be exposed by all adapters (*maintain existing interfaces*). Second, we use the adapter as an *intermediary* responsible for semantic translation of all the data received from different field devices into a standard format. As a side effect, using the adapter as an intermediary for semantic translation of data also addresses architectural driver 3, which requires support for converting nonstandard units by different devices.

The adapters are assigned the following functional responsibilities:

■ Send commands to the field device
■ Receive events from field device
■ Perform semantic translation for field device data

By virtue of addressing architectural drivers 1 and 3, adapters are assigned the following additional responsibilities:

■ Create standard interface to a device
■ Translate nonstandard units

Allocation of these responsibilities to the adapters leads to a realization that the BAS is still sensitive to a change in the number of field devices to which it is connected and must include logic to route commands and data to and from the correct adapter. To address this concern, we use the tactic of *hiding information,* introducing an adapter manager to hide information about the number and type of field devices actually connected. The adapter manager together with the adapters creates a virtual device; that is, for all other components of the building automation system, there is practically one field device to interact with at all times. The adapter manager internally maintains a routing table that associates logical identifiers with the field devices. The logical identifiers are associated with messages exchanged with BAS, and the adapter manager routes these messages to and from (adapters of) the correct field through the use of the routing table.

In addition, the adapter manager uses the following two architectural tactics to address deferring the binding time design concern:

- *Runtime registration:* This supports plug-and-play operation, allowing nonprogrammers to deploy new field devices.
- *Configuration files:* This tactic enables setting of configuration parameters, such as the type of a device and its initial property values, for bootstrapping a device at startup.

Essentially, when field engineers deploy a device, the adapter manager can read the configuration parameters from its configuration file, locate and deploy the correct adapter for the device, update the routing table, and start communicating with the device via its adapter.

The adapter manager is assigned the following functional responsibilities:

- Configure a field device
- Route data to a field device

By virtue of addressing architectural driver 1, the adapter manager is also assigned the following additional responsibilities:

- Defer binding to a field device until runtime
- Hide the number and type of field devices

The result of applying these tactics is shown in Figure 5.2.

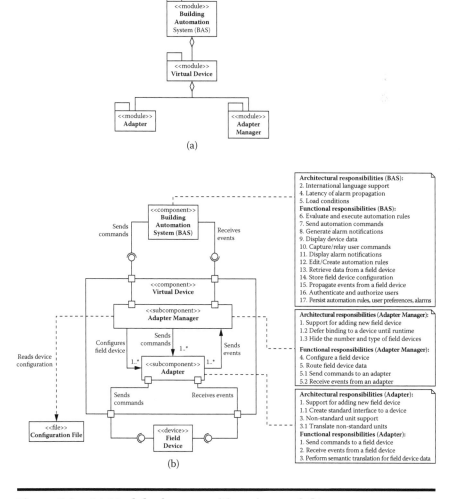

(a)

(b)

Figure 5.2   (a) Module decomposition view and (b) component-and-connector view showing building automation system after addressing support for adding new field devices.

Figure 5.2 shows how architectural driver 1 is satisfied partly by the adapter and partly by the adapter manager (by way of additional architectural responsibilities allocated to them), and together they implement what we refer to as a virtual device. BAS continues to act as a placeholder for the remaining architectural drivers and functional responsibilities that need to be addressed. Again, Figure 5.2a shows the static module decomposition and Figure 5.2b the dynamic component-and-connector view of the system, and we limit showing allocation of responsibilities to Figure 5.2b.

At this stage, architecture driver 1 is satisfied (and as a side effect, so is architecture driver 3), and we can move on to the next driver or set of drivers in the priority order. For instance, in this iteration we only treated a single driver (architecture driver 1) to drive the decomposition of the system, but this need not be the case. Drivers that have similar needs can be addressed together. Therefore, in the next iteration, we treat architecture drivers 4 and 5 together because they have similar performance concerns.

5.2.2 Addressing Latency and Load Conditions

Recall that driver 4 is concerned with regulatory compliance whereby a life-critical alarm has to be reported within 3 s of the occurrence of the event that raised the alarm. Driver 4, on the other hand, is related to a constraint that requires the building automation system to handle a wide array of configurations, ranging from 500 to 500,000 field devices. This constraint alerts us to the fact that a large number of field devices could create a significant load on the system, which must be managed to avoid an undesirable impact on how quickly an alarm event can make it through the system.

All of these requirements relate to the performance quality attribute that targets the latency and throughput needs of a system to make it more responsive. To address the latency and throughput needs of drivers 4 and 5, we can apply performance tactics further decomposing BAS to manage the demand for the resources and manage the resources more effectively so they can more efficiently meet this demand. Therefore, there are two design concerns to be addressed:

- *Manage demand for a resource*: The arrival of change-of-property-value events from the various field devices (for example, a thermostat reporting a change in its temperature property value) and the evaluation of automation rules in response to these events (for example, if the temperature reported by the thermostat is too high, then check the status of the fire alarms and smoke detectors) are sources of resource demands and must be managed.
- *Manage a resource*: The demand on resources may also have to be managed to reduce the latency of life-critical events and alarm propagation.

To address these concerns, we move the responsibility of rule evaluation and execution, and alarm generation, respectively, from BAS to a separate rule manager component and an alarm manager component. By virtue of being separate, these components can now be easily moved to dedicated execution nodes if necessary. In doing so, we are making use of the *increase available resources* tactic to address the resource management concern (alarm-handling capability obtains its own resources for efficient handling of life-critical events) and the *reduce computational overhead* tactic to address the resource demand concern (we can colocate all alarm-handling elements on their own dedicated node and reduce the latency of interelement communication).

We use an additional tactic to address the resource management concern. This tactic relies on introducing *concurrency* to reduce delays in processing time. Concurrency is used inside the rule manager and the alarm manager components to perform simultaneous rule evaluations. The results of applying these tactics are shown in Figure 5.3.

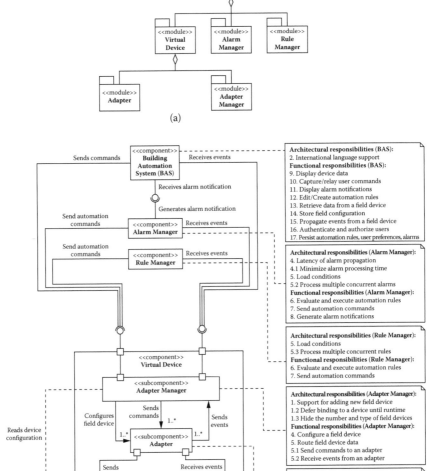

Figure 5.3 (a) Module decomposition view and (b) component-and-connector view showing building automation system after addressing support for performance architectural drivers.

As usual, we limit showing allocation of responsibilities to Figure 5.3b, and BAS acts as a placeholder for the remaining architectural drivers and functional responsibilities that need to be addressed. The following functional responsibilities have been assigned to the rule manager and alarm manager components:

■ Evaluate and execute automation rules
■ Send automation commands

By virtue of addressing driver 5, they both also obtain an additional architectural responsibility:

■ Process multiple concurrent rules

In addition, we assign the following functional responsibility to the alarm manager:

■ Generate alarm notifications

Because it also handles driver 4, we assign the following architectural responsibility to the alarm manager:

■ Minimize alarm-processing time

At this stage, architecture drivers 1, 3, 4, and 5 are satisfied, and only driver 2 remains. We address this driver in the next iteration.

5.2.3 *Addressing International Language Support*

Recall that driver 2 is concerned with a developer being able to package a version of the building automation system with new language support in 80 person-hours, giving users of the system the ability to personalize the user interface to a language and locale of their choice. These internationalization requirements relate to the modifiability quality attribute, the ease with which a system can be configured to support a new language and locale. To satisfy these requirements, we can apply the modifiability tactics to limit the impact of change and minimize the number of dependencies on the part of the system responsible for handling internationalization. There are two design concerns to be addressed:

■ *Localize changes*: This relates to minimizing the number of changes to be made to configure the building automation system for a new language and locale.
■ *Prevention of ripple effects*: This relates to minimizing the number of modules affected as a result of changes that result from configuring the building automation system for a new language and locale.

We address these concerns by further decomposing BAS into a separate presentation module and using the following modifiability tactics:

■ *Anticipation of expected changes*: Changes related to language and locale are localized to a single presentation module.
■ *Intermediary*: The presentation module also acts as an intermediary, preventing ripple effects of changes related to language and locale from propagating to the rest of the system.

We could have treated the building automation system as a collection of independently cooperating elements and chosen to associate a presentation manager with every module in the system, such as the alarm manager, rule manager, and virtual device. This would give us great flexibility in tailoring the user interface to the specific and unique human-computer interaction needs of a given module. However, the internationalization issues must be dealt with in multiple different modules. There is a trade-off between flexibility and modifiability. Because we do not need the added flexibility, we stick with the design decisions we have made. The results of applying these tactics are shown in Figure 5.4.

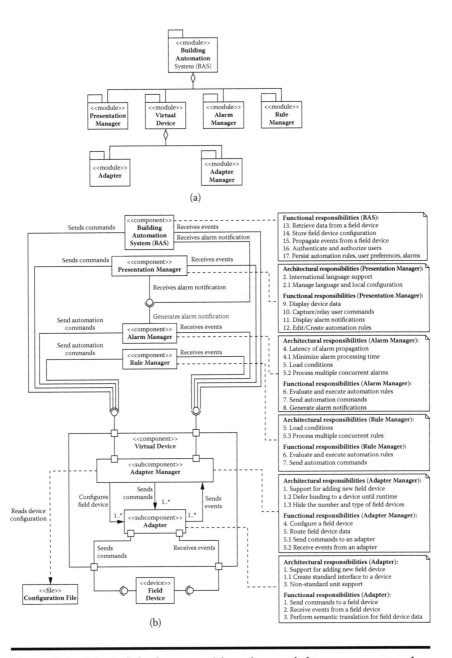

Figure 5.4 (a) Module decomposition view and (b) component-and-connector view showing building automation system after addressing support for international languages.

We assign the following responsibilities to the presentation module:

- Display device data
- Capture/relay user commands
- Display alarm conditions
- Edit/create automation rules

By virtue of addressing architectural driver 2, the presentation module receives the following additional architectural responsibility:

- Manage language and locale configuration

5.3 Architecture Trade-Offs

As previously mentioned, the architecture elaboration process we are using is iterative. Moreover, the tactics we choose to implement can often have a negative impact on the quality attributes they do not target specifically. In the case of the building automation system, we focused on modifiability and performance tactics, which can have a negative impact on each other. We revisit the modifiability and performance drivers next to address these issues.

5.3.1 Revisiting Modifiability Drivers

Introducing performance tactics resulted in creation of multiple components (rule manager and alarm manager, for instance) that now depend on the virtual device. Therefore, based on its current structure, we can predict that some changes to the virtual device have the potential to propagate to several other components of the building automation system. We would like to minimize the ripple effect of these changes. To achieve this objective, we introduce a publish-subscribe bus as shown in Figure 5.5.

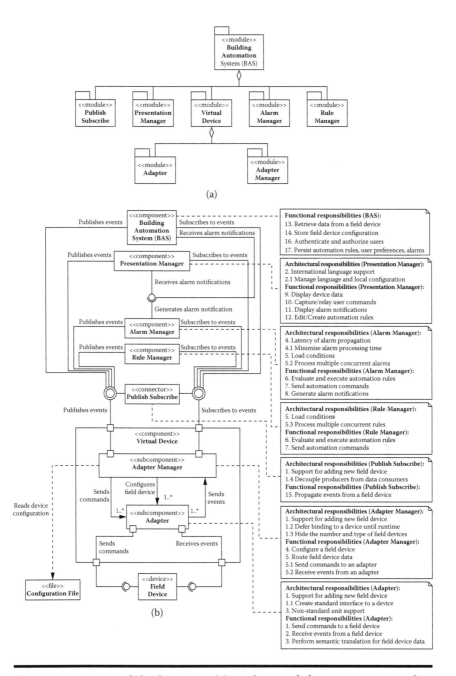

Figure 5.5 **(a) Module decomposition view and (b) component-and-connector view showing building automation system after addressing the modifiability and performance trade-off.**

The publish-subscribe bus uses three modifiability tactics. First, it alleviates the syntactic dependencies of intercomponent calls by acting as a standard interface *intermediary*. Second, using the *module generalization* tactic makes it invariant to the type of events it transports. This generalization allows new types of events to be transported with no modification to the publish-subscribe component. Finally, it relies on *runtime registration* to allow system extensibility by adding publishers and subscribers.

We allocate the following responsibility to the publish-subscribe bus:

■ Propagate change-of-value notifications

By virtue of decoupling the rest of the system from the virtual device, we also assign to it the following additional architectural responsibility:

■ Decouple data consumers from data producers

5.3.2 Revisiting Performance Drivers

By examining the current system structure, it can be seen that every time the rule manager, the alarm manager, or the presentation manager needs to query a field device, it needs to make a call that traverses multiple components along the way to the field device. Because crossing component boundaries typically introduces computational overhead, and because the querying latency of field devices is a constraint over which we have no control, we decompose the virtual device and introduce a cache component to improve device-querying performance. This is shown in Figure 5.6.

This cache provides field device property values to the system, saving part of the performance cost incurred when querying the actual field devices. The performance gains are seen because we reduce the number of component and

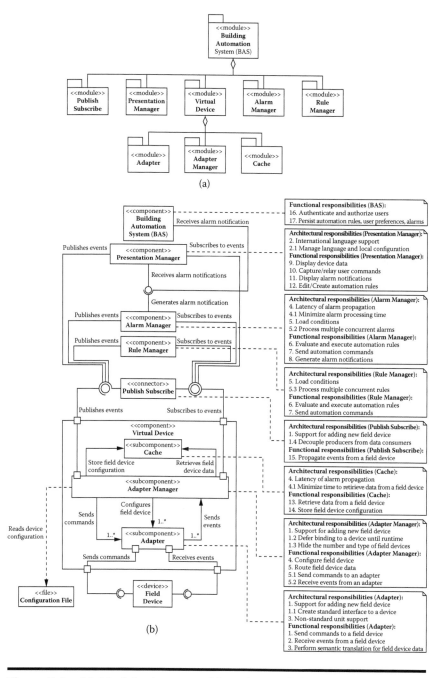

Figure 5.6   (a) Module decomposition view and (b) component-and-connector view showing building automation system after addressing the performance and modifiability trade-off.

machine boundaries traversed for each query. A cache is really the application of the performance tactic of *the maintaining multiple copies of data*. We allocate the following functional responsibilities to the cache:

- Retrieve data from a field device
- Store field device configuration

In addition, we assign the following architectural responsibility related to improving performance:

- Minimize time to retrieve data from a field device

> There is never a perfect architecture that optimally satisfies all of its requirements. All architectural designs involve trade-offs.

5.4 The Final Architecture

At this stage, all architectural drivers have been addressed, but BAS, which has been acting as a placeholder for unaddressed requirements, still has two functional requirements related to managing authentication/authorization of users of the system and the persistence of rules, alarms, and user preferences. We introduce a new component, BAS server, assigning it these remaining responsibilities. The final architecture is shown in Figure 5.7.

5.5 Summary

This chapter explored the shaded portion of International Organization for Standardization/International Electrotechnical

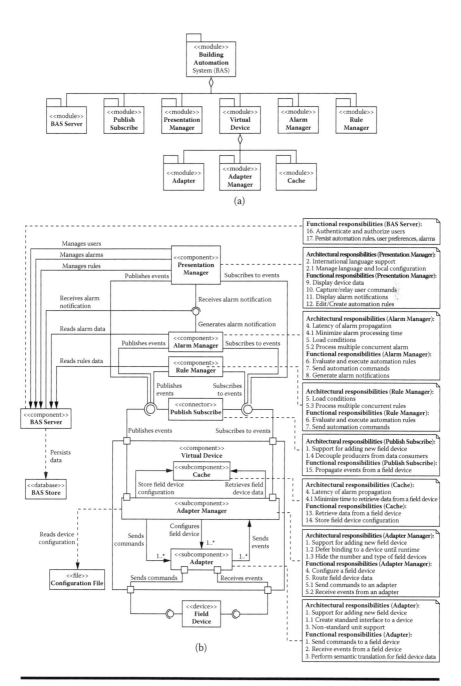

Figure 5.7   (a) Module decomposition view and (b) component-and-connector view showing building automation system after addressing all remaining requirements.

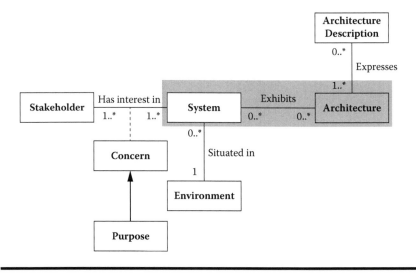

Figure 5.8   Conceptual framework for system and software architecture.

Commission (ISO/IEC) 42010 conceptual framework for system and software architecture shown in Figure 5.8.

We demonstrated the use of ADD, a recursive decomposition method for architecture design, by applying it to the building automation system. We used numerous patterns and tactics that were discussed in the previous chapter to address design concerns inherent in the architectural requirements that drove the design of this system. In the process, we saw how architectural drivers conflict with each other and create trade-off situations that need to be resolved. In the end, the final architecture is not one that was perfect but one that tries to achieve a balance among competing design forces, some of which may conflict with each other.

5.6 Questions

1. Refine the architecture presented in Figure 5.7 to improve the availability of the *alarm component*. Appropriately

allocate functional responsibilities and architectural drivers to the elements of the resulting architecture.

2. Tactics and patterns are important design decisions an architect makes to satisfy quality attribute requirements of a system. When analyzing a design or architecture of a system, the presence of these design decisions gives some measure of confidence that a certain quality attribute requirement will be satisfied. Justify your choices of the patterns or tactics for improving the availability of the alarm component in problem 1.

3. Suppose a virtual field device needs to be highly available, with the stakeholder-provided availability scenario described in the following table:

Stimulus	A virtual field device hardware or software component fails.
Stimulus source	A fault occurs in a virtual field device hardware or software component.
Environment	At the time of failure, a virtual field device may be servicing a number of requests concurrently with other queued requests.
Artifact	Virtual field device.
Response	All query requests made before and during failure must be honored.
Response measure	The processing of requests must resume within a second.

A few of the design concerns related to this scenario are shown in the following table:

Design Concerns	Subordinate Concerns	Description
Fault detection	Health monitoring	Detect fault and notify an element to deal with the fault

Design Concerns	Subordinate Concerns	Description
Fault preparation	Restart	Restart a component when failure occurs
	Data integrity	Ensure that when a failure occurs there is sufficient state information on the failed component for recovery to occur
Fault recovery	Transparency to clients	Fault must be transparent to the clients

    a. List alternative patterns/tactics that can be used to address these concerns.

    b. Select patterns from the list and provide reasoning behind your choice.

    c. Capture your solution in an architecture view.

    d. Allocate responsibilities to the different elements in your architecture.

    e. Create a timing model to support your solution.

4. A patient-monitoring system is a real-time embedded system used in hospitals to monitor a patient's vital signs. It is a stand-alone bedside unit that obtains a patient's data via pods attached to the patient or to other wide range of monitoring devices and communicates via a network to a central station. When the patient is being transported, the base unit can be detached from its docking station, and it travels with the patient. The base unit has a display that shows a patient's physiological data as waveforms, parameter values, and alarms. For example, heart rate and pressures are discrete intermittent parameter values, whereas an electrocardiographic (EKG) recording of heart activity or arterial pressure is a continuous waveform sample. Hospital personnel select the kind of data to be displayed and display format. They can configure the alarms to go

off when values fall outside the range they specify. The system can also display as much as 24 h of trend data. The product supports hundreds of customized user settings and thousands of application policy rules. These policy rules apply to features like EKG processing, parameter processing, alarms processing, patient admission, patient setup, display of patient information, two-channel local recording, network support, and acquisition and display of waveform and parameter information from other devices, such as ventilators and anesthesia systems. Some significant business goals for the patient-monitoring system, their associated engineering objectives, quality attributes, and quality attribute scenarios along with the priorities are given in the following table below; using this information as input, create an architecture of the patient-monitoring system using the attribute-driven design method:

Business Goal	Engineering Objective	Quality Attribute	Quality Attribute Scenario	Priority
Support customized user settings and application policy rules	Support displaying measurements from a wide variety of devices	Modifiability Availability	A new patient-monitoring device is added to the system and configured at runtime with no downtime or side effects.	(H, H)
	Configure system display for customized output	Usability	An end user requests information from a patient-monitoring device. The system automatically configures its display to show the data in the requested format.	(H, H)
Minimize downtime and failure	Support display of correct information 24 h a day and 7 days a week	Availability	The system is disconnected from its docking station when transporting a patient. The system continues to operate with no downtime.	(H, H)

Note: H, high.

References

L. Bass, P. Clements, and R. Kazman, *Software Architecture in Practice*, third edition. Boston: Addison-Wesley, 2013.

R. Sangwan, C. Neill, M. Bass, and Z. El Houda, Integrating a software architecture-centric method into object-oriented analysis and design. *J. Syst. Softw.* 81, 727–746, 2008.

W. Woods, *A Practical Example of Applying Attribute-Driven Design (ADD)*, version 2.0 (CMU/SEI-2007-TR-005). Pittsburgh, PA: Carnegie Mellon Software Engineering Institute.

Chapter 6

Communicating the Architecture

6.1 Introduction

Architecture is an important artifact that serves many different purposes. Project managers use it to organize a project and distribute work among its teams. They also use it periodically for educating new members who join the project. Project teams use the architecture as a blueprint for development and reasoning about the system. Once the system is operational, maintainers use architecture for system understanding and maintenance. Not only is communicating an architecture critical but also we may need to carefully construct its documentation to cater to a diverse set of stakeholders that may benefit from it.

6.2 Views as a Basis for Documentation

So far, architecture has been described as an organization of a system into its constituent elements and the relationship among these elements. Module decomposition and component-and-connector diagrams have been predominantly used to document this organization. These diagrams, however, are just one of the many different views of the architecture.

Which views you document depends on the uses you expect to make of the documentation. For instance, developers are concerned with how a system is structured so they can understand how elements they are developing depend on the rest of the system and how these elements interact with each other at runtime. A module view can be used to depict the static development time structure of the system, showing its decomposition into different modules along with their interdependencies. A component-and-connector view, on the other hand, can be used to depict the dynamic runtime structure of the system, showing its decomposition into elements (such as components, tasks, processes, threads, etc.) that have runtime behavior and interaction. The module view can be used for reasoning about modifiability concerns and allows developers to understand how changes they make can have an impact on others and how changes others make have an impact on them. The component-and-connector view can be used for addressing concerns related to many of the runtime qualities of a system, such as performance, security, and availability.

The best way to document an architecture is to enumerate its stakeholders, understand their concerns, and document views that address these concerns (Clements et al., 2010). In this way, you will be creating an architecture documentation that can be selectively and effectively navigated by those interested in it. One need not read the document cover to cover, which is often the case in poorly structured architecture description documents, which frequently are hundreds of pages long.

6.3 Documenting a View

A view's primary function is to show the structure that it represents. Its documentation, therefore, mainly consists of the following:

- *Primary presentation*: a visual representation that shows the elements in a structure and the relationships among them.
- *Element catalog*: details at least those elements and relations depicted in the primary presentation; these details include the interfaces of the elements and how these elements behave at runtime.
- *Context diagram*: shows how the system depicted in the view relates to its environment.
- *Architecture background*: explains the design rationale, analysis results, and assumptions.

There is some documentation that applies to the entire architecture and to all the views, including

- Information on how the documentation is organized:
  - *Documentation road map*: describes what the views are, their intended purpose, and where they can be found in the document
  - *View template*: describes standard organization of a view
- Information on what the architecture is:
  - *System overview*: short description of what the system function is, who its users are, and any important background or constraints
  - *Mapping between views*: relationship among views showing how the architecture works as a unified conceptual whole
  - *Element list*: index of all of the elements that appear in any of the views along with a pointer to where each one is defined

  – *Project glossary*: defines terms unique to the system with special meaning
- Information on why the architecture is the way it is:
  – *Rationale*: explains how the overall architecture is a solution to its requirements, why it was chosen, and the implications in changing it

6.4 Building an Architecture Description Document

The views-and-beyond approach (Clements et al., 2010) as outlined in the previous section suggests the following structure:

- Section 1: Document Road Map
- Section 2: System Overview
- Section 3: View Template
- Section 4: Views
  – Section 4.1 View 1
  – Section 4.2 View 2
  – . . .
  – Section 4.*n* View *n*
- Section 5: Mapping between Views
- Section 6: Element List
- Section 7: Project Glossary
- Section 8: Rationale

If a system is rather large, an alternative is a multivolume documentation in which Volume 1 would consist of the documentation that applies to all the views (Sections 1–3, 5–8), and Volume 2 would consist of all the views (Section 4 and its subsections). Each view can also be documented in a separate volume if desired.

In the following sections, the building automation system is used as an illustrative example to demonstrate the views-and-beyond approach for creating an architecture description document.

6.5 Architecture Description for the Building Automation System

The structure of the document follows the views-and-beyond approach just described. Each document section is labeled as Section 1, Section 1.1, Section 1.2, Section 2, and so on (do not confuse them with the Chapter 6 section numbering, which begins with "6.": Section 6.5.1, Section 6.5.1.1, etc.). One should go through these sections as though reading an independent document submitted by an organization chartered with the development of the building automation system.

6.5.1 Section 1: Document Road Map

The document road map gives an overview of the structure and content of the entire architecture description document and provides guidance on how different stakeholders might want to view the various sections.

6.5.1.1 Section 1.1: Description of the Architecture Documentation

The architecture description of the Building Automation System (BAS) contains the following sections:

Section 1, Architecture Documentation Road Map, lists and outlines the contents of the overall documentation package, including how the various sections address different stakeholder concerns. This is the first section that a new stakeholder must read.

Section 2, System Overview, enumerates the business goals and engineering objectives motivating this project and gives a description of the system's operational context, a broad set of requirements (functional and quality attributes) it must satisfy, and some of the major design constraints. The purpose is to help the reader understand what the architecture is trying to achieve.

Section 3, Architectural View Template, explains how the architectural views are described in the document. The purpose is to help the reader understand the information provided for each of the architectural views in this document.

Section 4, Software Architecture Views, describes a number of different views of the architecture, with each view addressing specific stakeholder concerns.

Section 5, Mapping between Views, draws comparisons between different views of the system, highlighting areas that are common to these views.

Section 6, Rationale, explains the rationale behind the architecture for the overall system.

6.5.1.2 Section 1.2: How Stakeholders Can Use the Documentation

All stakeholders must read the system overview section to understand the business goals motivating this project and to gain a broad understanding of the requirements driving the development of BAS. In addition, the following list enumerates the different stakeholder roles of primary importance to the system and describes, for each stakeholder role, the sections of this document that might be most relevant for addressing their concerns:

Someone new to the project: Read the architecture documentation road map section for an overview of the structure and content of the architecture documentation and decide which sections of the document are most relevant to address the concerns for your assigned role. Read the view template section to understand how each view of the architecture is documented before studying the views most relevant to your role.

Project manager: The module view can provide an overview of the structure of the project; the component-and-connector view can help identify components that will have to be qualified, procured, and integrated; and the deployment view will provide information on the hardware environment to be acquired and the testing environment to be set up.

Software architect: The module view can help reason about modifiability and in understanding the impact of change. The component-and-connector views help reason about runtime quality attribute requirements, such as performance, security, and availability. They can also be used to infer progression of data through the system and runtime behavior of the system.

Performance engineer: The component-and-connector views can be used for understanding the behavior of the system. The deployment view can be used to understand how software is allocated to hardware.

Security analyst: The component-and-connector views can be used for understanding the interaction among the runtime components of BAS. The deployment view provides information on the physical environment in which these components operate.

Developer: The module view can help the developer understand intermodule dependency and the structure of the project. The component-and-connector views can help infer progression of data through the system and runtime behavior of the system.

Maintainer: The module view can serve as the basis for impact analysis, that is, which module must be changed to address a maintenance issue and what other modules are likely to be impacted by this change. The component-and-connector views should provide information on the software components that exist and their responsibilities. Together, the information in these views can be useful as a starting point for future extensions or modifications.

Customer/acquirer: The component-and-connector and the deployment views can help gain an understanding of how the system is structured to carry out its mission and to gain an appreciation for the work that must be done to build it.

Users: The component-and-connector views should provide basic understanding for the structure of the system and how its various parts behave.

6.5.2 Section 2: System Overview

The company manufactures devices for the building automation domain and software applications that manage a network of these devices. With the hardware being commoditized, its profit margins have been shrinking. The internal development costs for the software applications that manage different devices have also been rising. To sustain their business long term, the company has decided to create a new integrated building automation system.

6.5.2.1 Section 2.1: Business Goals

The business goals in undertaking this development are as follows:

- *Reduce development costs*: Consolidate all existing software systems for managing hardware devices for building automation into a single unified building automation system.
- *Create new revenue streams*: Charge existing customers for the management system, enter new and emerging geographic markets, and open new sales channels by partnering with value-added resellers (VARs).

Table 6.1   Business Goals and Engineering Objectives

Business Goal (Mission Objective)	Goal Refinement (Engineering Objective)
Expand by entering new and emerging geographic markets	Support international languages
	Comply with regulations that have an impact on life-critical systems, such as fire alarms, to operate within specific latency constraints
Open new sales channels in the form of value-added resellers (VARs)	Support hardware devices from different manufacturers
	Support conversions of nonstandard units used by the different hardware devices

Table 6.1 captures these business goals and their refinement into engineering objectives for the design of the building automation system.

The reduction of internal development costs can be achieved by aggregating the development and maintenance functions of the several small applications into one project for the building automation system. It has more to do with changes to the organizational structure than the architecture of the system to be developed and therefore is not included in Table 6.1.

6.5.2.2 Section 2.2: System Context

Figure 6.1 gives an operational view of the building automation system. The overall goal of the building automation system is to monitor the locations within the buildings (using field devices) for normal building operations such as HVAC (heating, ventilation, air conditioning), lighting, safety, and security. These devices periodically report back the conditions, and the facility managers can monitor these conditions on their management workstations. Conditions that may threaten the safety and security of the building residents are raised as alarms so

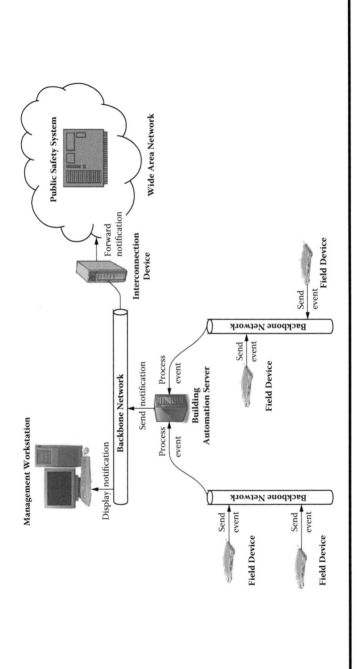

Figure 6.1 Operational view.

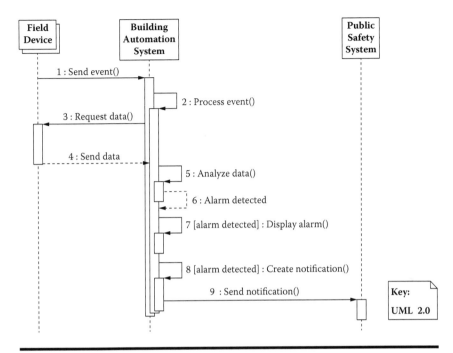

Figure 6.2 Operational scenario showing major flow of events.

that prompt action can be taken. Figure 6.2 shows one such operational scenario.

As shown in Figure 6.2, many field devices concurrently transmit events to the building automation system. The building automation system processes these events, requesting additional data from these devices to determine if there is a safety hazard. If so, an alarm is generated; it is displayed on a management workstation, and a notification is sent to the public safety system.

In addition to facilities managers monitoring the building for normal operations, field engineers intend to manage the field devices and dynamically reconfigure them, and the system administrators intend to manage the users of the building automation system. The overall context of the building automation system is shown in Figure 6.3.

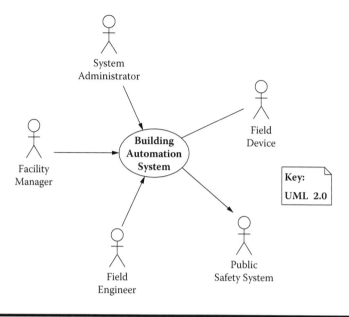

Figure 6.3 System context.

6.5.2.3 Section 2.3: Functions

The building automation system must provide certain critical features to support the engineering objectives shown in Table 6.1. For instance, integration implies the features of existing applications to be integrated must be supported in the new system. This may require innovative ways of displaying information in the user interface and providing fine-grained access control on who is allowed to interact with what part of the system.

Supporting international languages implies personalization capabilities. Regulatory policies for safety-critical parts of the system would require alarm-handling capabilities for situations that could cause loss of life. Supporting hardware devices from different manufacturers would require dynamic configuration capabilities.

Figure 6.4 shows the refinement of these features into specific use cases based on how the external actors shown in the context diagram in Figure 6.3 intend to use the system.

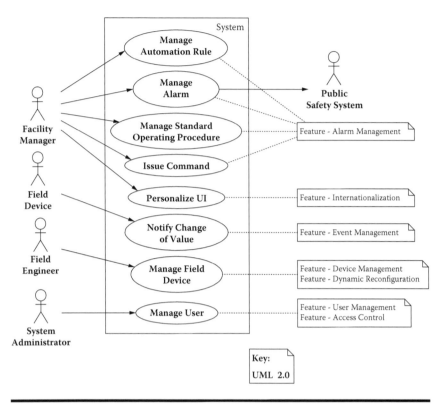

Figure 6.4 Use cases.

The use cases imply the following functional responsibilities that the building automation system must fulfill:

1. Send commands to a field device
2. Receive events from a field device
3. Perform semantic translation for field device data
4. Configure a field device
5. Route data to a field device
6. Evaluate and execute automation rules
7. Send automation commands
8. Generate alarm notifications
9. Display device data
10. Capture/relay user commands
11. Display alarm notifications
12. Edit/create automation rules

13. Retrieve data from a field device
14. Store field device configuration
15. Propagate change-of-value notifications
16. Authenticate and authorize users
17. Persist automation rules, user preferences, alarms

6.5.2.4 Section 2.4: Quality Attribute Requirements

Critical qualities the building automation system must exhibit to support the engineering objectives are shown in Table 6.2. Each

Table 6.2 Quality Attributes and Scenarios Derived from Engineering Objectives

Engineering Objective	Quality Attribute	Quality Attribute Scenario
Support hardware devices from many different manufacturers	Modifiability	A field engineer is able to integrate a new field device into the system at runtime and the system continues to operate with no downtime or side effects.
Support conversions of nonstandard units used by the different devices	Modifiability	A system administrator configures the system at runtime to handle the units from a newly plugged-in field device and the system continues to operate with no downtime or side effects.
Support international languages	Modifiability	A developer is able to package a version of the system with new language support in 80 person-hours.
Comply with regulations requiring life-critical systems to operate within specific latency constraints	Performance	A life-critical alarm should be reported to the concerned users within 3 s of the occurrence of the event that generated the alarm.

quality attribute is elaborated with a concrete scenario that can be used to assess the suitability of the final architecture.

6.5.2.5 Section 2.5: Constraints

Table 6.3 enumerates additional factors that have to be considered as they may constrain how the architecture for the building automation system is designed.

6.5.2.6 Section 2.6: Architectural Drivers

A prioritized list of architectural drivers for the system is shown in Table 6.4 along with the sources (quality attribute scenarios, product features, constraints) from which they were derived. The priorities are shown as tuples derived by soliciting input from both the business and the technical stakeholders. The business stakeholders prioritized drivers based on their business value (H, high, implies that a system lacking that capability will not sell; M, medium, implies that a system lacking that capability will not be competitive; L, low, implies it is something nice to have), whereas the technical stakeholders did so based on how difficult it would be to achieve a given scenario during the system design (H, high, implies a driver that has a systemwide

Table 6.3 Constraints

Category	Factor	Description
Organization	Expertise	The development organization has a strong background in development on the Microsoft platform.
Technology	Cost	The system must reuse the company's existing rules engine.
Product	Variability	The system must handle a wide range of disparate field devices and configurations, ranging from 500 to 50,000 such devices, at customer sites.

Table 6.4 Architectural Drivers

No.	Architectural Driver	Source	Priority
1	Support for adding new field device	• *Quality attribute scenario*: Support hardware devices from many different manufacturers • *Product feature*: Device management, dynamic reconfiguration • *Constraint*: Product variability	(H, H)
2	International language support	• *Quality attribute scenario*: Support international languages • *Product feature*: Internationalization	(M, M)
3	Nonstandard unit support	• *Quality attribute scenario*: Support conversion of nonstandard units used by different devices • *Product feature*: Device management, dynamic reconfiguration • *Constraint*: Product variability	(H, M)
4	Latency of alarm propagation	• *Quality attribute scenario*: Comply with regulations • *Product feature*: Event management, alarm management	(H, H)
5	Load conditions	• *Constraint*: Product variability	(H, H)

Note: H, high; M, medium.

impact; M, medium, implies a driver that has an impact on a significant part of the system; L, low, implies a driver whose impact is fairly local and therefore not architecturally significant). This results in nine different combinations in the following order of precedence: HH, HM, HL, MH, MM, ML, LH, LM, and LL.

Architectural drivers 1 through 5 relate to the quality attribute scenarios enumerated in Table 6.2. In addition, architectural drivers 1 and 3 also correspond to the device management and dynamic reconfiguration, 2 to the internationalization and localization, and 4 to event management and alarm management product features shown in Figure 6.4. Also, architectural drivers 1, 3, and 5 address the product variability constraint from Table 6.3.

6.5.3 Section 3: View Template

A view's primary function is to show the structure that it represents. Its documentation, therefore, consists of the following:

- *Primary presentation*: shows the elements in a structure and the relationships among them.
- *Element catalog*: details at least those elements and relations depicted in the primary presentation; these details include the interfaces of the elements and how these elements behave at runtime.
- *Architecture background*: explains the design rationale, analysis results, and assumptions.

All views contained in Section 4 use this standard template.

6.5.4 Section 4: Views

The purpose of each view is to address specific stakeholder concerns as discussed previously.

6.5.4.1 Section 4.1: Module View

The module view serves as the basis for making decisions on how to structure the system into cohesive units of implementation such that

- The impact of an anticipated change is localized;
- Making this change does not ripple to other parts of the system that depend on the element being changed;

■ When desired, the change can be made while the system is in operation without any downtime or side effect.

6.5.4.1.1 Section 4.1.1: Primary Presentation

Figure 6.5 shows the major modules that make up the building automation system.

6.5.4.1.2 Section 4.1.2: Element Catalog (Modules)

Table 6.5 lists the responsibilities of the major modules in the building automation system.

6.5.4.1.3 Section 4.1.3: Architecture Background

Figure 6.6 highlights the design decisions that were made to improve the modifiability of the building automation system.

The design decisions, marked in Figure 6.6 and cross-referenced in the material that follows, achieve the following:

■ **Decision 1 (virtual device)**: Hides communication with the field devices
■ **Decision 2 (presentation manager)**: Hides details of user interaction, their language, and locale preferences
■ **Decision 3 (publish-subscribe)**: Decouples data providers from data consumers, providing a generalized interface for reliable event delivery

Together, these decisions satisfy architecture drivers 1–3 (see Table 6.4).

6.5.4.2 Section 4.2: Component-and-Connector View

The component-and-connector view serves as the basis for making decisions on how to structure the system into runtime units of interaction such that

■ Resources manage their demand efficiently, minimizing latency and maximizing throughput.

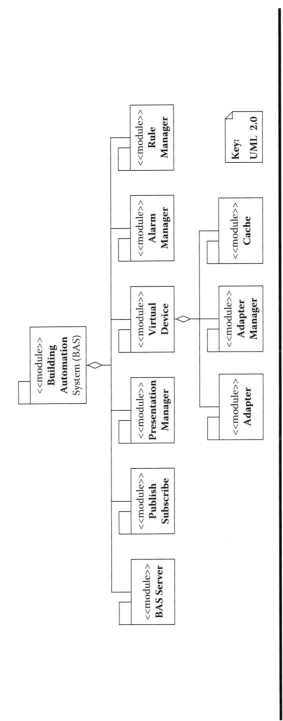

Figure 6.5   Module decomposition view.

Table 6.5 Major Modules and Their Responsibilities

	Module	Responsibilities
1	Adapter	1. Create standard interface to a field device 2. Translate nonstandard units 3. Send commands to a field device 4. Receive events from a field device 5. Perform semantic translation for field device data
2	Adapter manager	1. Defer binding to a device until runtime 2. Hide the number and types of field devices 3. Configure a field device 4. Route field device data
3	Alarm manager	1. Minimize alarm-processing time 2. Process multiple concurrent alarms 3. Evaluate and execute automation rules 4. Send automation commands 5. Generate alarm notifications
4	BAS server	1. Authenticate and authorize users 2. Persist automation rules, user preferences, and alarms
5	Cache	1. Minimize time to retrieve data from a field device 2. Retrieve data from a field device 3. Store field device configuration
6	Presentation manager	1. Manage language and locale configuration 2. Display device data 3. Capture/relay user commands 4. Display alarm notifications 5. Edit/create automation rules
7	Publish-subscribe	1. Decouple data producers from data consumers 2. Propagate events from a field device
8	Rule manager	1. Process multiple concurrent rules 2. Evaluate and execute automation rules 3. Send automation commands

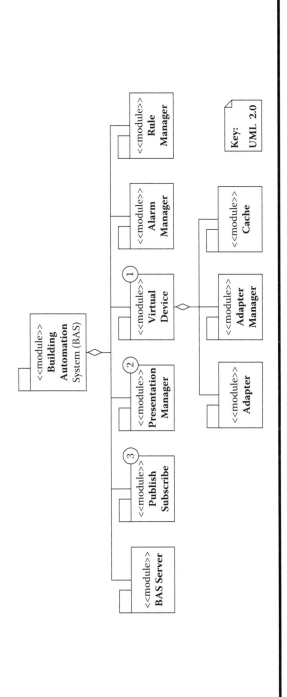

Figure 6.6 Module view with design decisions highlighted.

6.5.4.2.1 Section 4.2.1: Primary Presentation

Figure 6.7 shows the major components that make up the building automation system.

6.5.4.2.2 Section 4.2.2: Element Catalog (Components)

Table 6.6 lists the responsibilities of the major components in the building automation system.

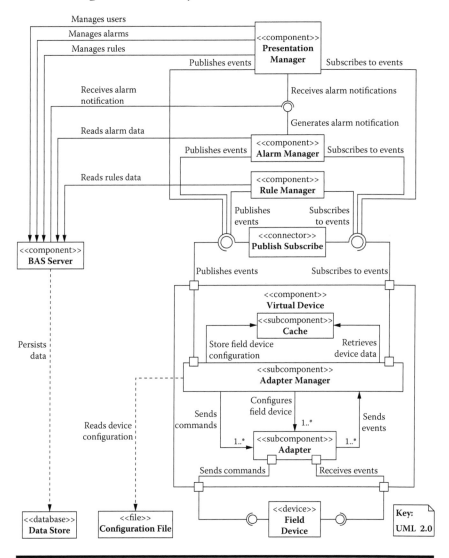

Figure 6.7   Component-and-connector view.

Table 6.6 Major Components and Their Responsibilities

Component	Responsibilities
Alarm manager	• This component handles the triggering alarm rules and sending notifications when alarms are generated.
BAS server	• This controls all user access to the building automation system and its data along with persistence of this data.
Presentation manager	• This supports a browser for viewing field device property values, an alarm viewer, a rules editor for rules manager and alarm rules, and an interface for issuing commands.
Rule manager	• This component handles triggering automated rules and commanding field devices in response to desired actions of the triggered rules.
Virtual device	• This handles configuration of, routing commands to and receiving data from, field devices, including any semantic translation of data.

6.5.4.2.3 Section 4.2.3: Element Catalog (Connectors)

Table 6.7 lists the responsibility of the major connector in the building automation system.

6.5.4.2.4 Section 4.2.4: Element Catalog (Behavior)

Figure 6.8 shows a process view that captures how components behave at runtime. Many field devices concurrently transmit data to the virtual field device. The virtual field device processes the raw data, which are then made available for analysis to the alarm subsystem via the publish-subscribe

Table 6.7 Major Connector and Its Responsibility

Connector	Responsibility
Publish-subscribe	• Receives events from virtual device and propagates them to subscribers of these events

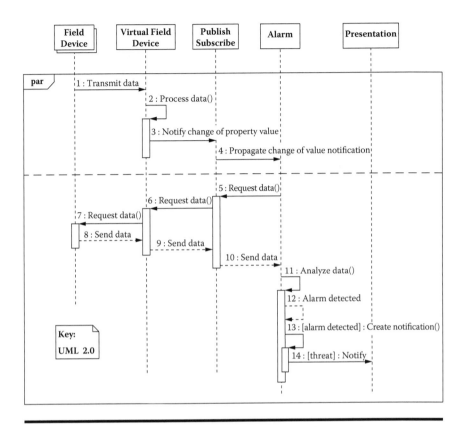

Figure 6.8 A process view showing component interactions.

connector. When alarms are detected, the alarm manager notifies the presentation component, which displays the alarm on a facilities manager's workstation.

6.5.4.2.5 Section 4.2.5: Architecture Background

Figure 6.9 highlights the design decisions that were made to improve the performance of the building automation system. The design decisions, marked in the figure and cross-referenced here, achieve the following:

- **Decision 1 (alarm manager)**: Minimizes alarm propagation latency through concurrent processing of alarm events and concurrent evaluation of alarm rules

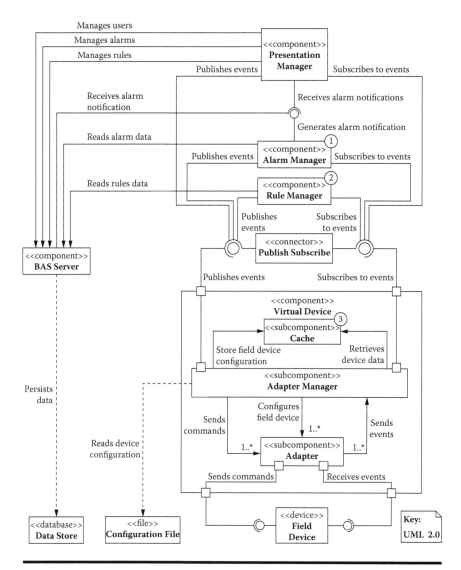

Figure 6.9 Component-and-connector view with design decisions highlighted.

■ **Decision 2 (rule manager)**: Reduces load by offloading from the system, processing of field device events and their related rules

■ **Decision 3 (cache)**: Reduces latency of retrieving property values from a field device

Together, these decisions satisfy architecture drivers 4 and 5 (see Table 6.4).

6.5.4.2.5.1 Section 4.2.5.1: Performance Model—We have created analytic models (along with prototypes/simulations) to have further confidence in the design decisions we have made. Figure 6.8 shows message flow across key hardware and software elements; in the event that data from the field devices begin to indicate the possibility of an alarm, the facilities managers need to know about this possibility within 3 s of its occurrence (architecture driver 4 in Table 6.4).

Table 6.8 shows total computer resource requirements for this model. We have made a simplifying assumption that all components are deployed and running together on the same machine; hence, the time taken by the publish-subscribe to propagate data received from the virtual device to the alarm subsystem is negligible. It is therefore ignored in subsequent calculations.

If a central processing unit (CPU) takes 0.00001 s to execute 1,000 instructions, a disk takes 0.02 s for every physical input/output (I/O), and network message takes 0.01 s processing time, the best-case elapsed time for the alarm detection scenario is

Table 6.8   Computer Resource Requirements for Alarm Processing

Processing Step	CPU Instructions (K)	Physical I/O	Network Messages
Transmit data (field device)	30	6	3
Process data (virtual field device)	60	8	4
Analyze data (alarm)	130	6	3
Total	220	20	10

Note: I/O, input/output.

$$(220 \times 0.00001) + (20 \times 0.02) + (10 \times 0.01) = 0.50220 \text{ s}$$

This is well under our 3-s requirement, ignoring network delays and processing overhead that may result when components are deployed on different machines.

6.5.4.3 Section 4.3: Deployment View

The deployment view serves as the basis for making decisions on how to structure the system into runtime units of interaction such that

- Components can be colocated and deployed together to reduce interprocess communication overhead.
- Components can be deployed independently and resources can be added seamlessly for scaling the system to meet its resource demand.

6.5.4.3.1 Section 4.3.1: Primary Presentation

Figure 6.10 shows the deployment and interactions of the major components that make up the building automation system.

6.5.4.3.2 Section 4.3.2: Element Catalog

Table 6.9 lists the responsibilities of the major components deployed in the building automation system.

6.5.4.3.3 Section 4.3.3: Architecture Background

Figure 6.11 highlights the design decisions that were made to improve the scalability of the building automation system. The design decisions (marked 1 through 4) in the figure provide a great deal of flexibility in how components are deployed. They allow the marked components to be deployed independently as a .NET assembly and allocated their own resources, thus distributing system load and improving scalability. By the same

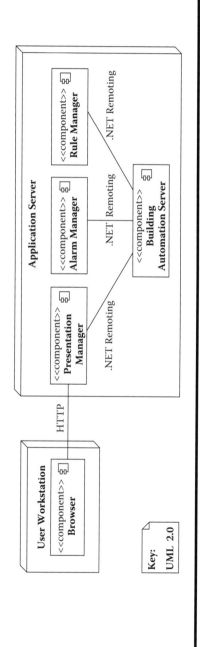

Figure 6.10 Deployment view.

Table 6.9   Major Deployment Units and Their Responsibilities

Element	Description
Alarm manager	The alarm manager runs as a separate process and will communicate, using .NET remoting, with the various components for managing the life cycle of an alarm.
BAS server	Deployed as a .NET executable in a separate process, this is the main component that controls user access to the building automation system and manages event and data flow to/from the field devices and other components using .NET remoting.
Browser	There will be some code in the form of Javascript in the browser used for interaction with the presentation manager for displaying alarms and status of different field devices on the user workstation.
Presentation manager	This component contains one or more .NET assemblies for the web page and other client-side resources that will be executed as a dedicated ASP .NET process and will communicate with other components by means of .NET remoting.
Rule manager	The rule manager runs as a separate process and communicates, using .NET remoting, with the various components when evaluating rules and commanding field devices.

token, some of them could be colocated on a single machine if desired and execute as a separate process or the same process to reduce communication overhead caused by crossing machine and network boundaries. Together, these decisions satisfy performance-related architecture driver 5 (see Table 6.4).

6.5.5 Section 5: Mapping between Views

Mapping between views draws comparisons between different views of the system, highlighting areas that are common to

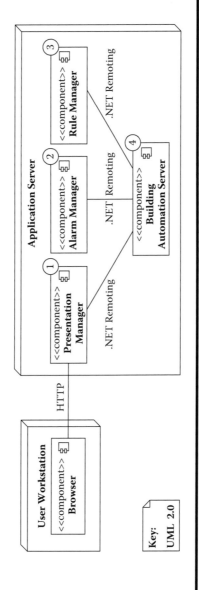

Figure 6.11   Deployment view with design decisions highlighted.

Table 6.10   Mapping between Module and Component-and-Connector (C&C) Views

Module View	C&C View	Relation
Adapter	Virtual device	The named modules are packaged together as a part of the mapped component.
Adapter manager		
Cache		
Alarm manager	Alarm manager	
BAS server	BAS server	
Presentation manager	Presentation manager	
Publish-subscribe	Publish-subscribe	
Rule manager	Rule manager	

these views. Table 6.10 shows how modules map to the different components of the building automation systems and the relationship among them. Table 6.11 shows how components in the component-and-connector view map to those deployed at runtime and the relationship among them.

Table 6.11   Mapping between Component-and-Connector (C&C) and Deployment Views

C&C View	Deployment View	Relation
Publish-subscribe	BAS server	The named C&C elements are part of the deployed component.
Virtual device		
BAS server		
Alarm manager	Alarm manager	
Rule manager	Rule manager	
Presentation manager	Presentation manager	

6.5.6 *Section 6: Rationale*

This section explains how the overall architecture is a solution to its requirements, why it was chosen, and the implications in changing it:

> *Business context*: Our company manufactures devices for the building automation domain and software applications that manage a network of these devices. With the hardware being commoditized, its profit margins have been shrinking. The internal development costs for the software applications that manage different devices have also been rising. To sustain their business long term, the company has decided to create a new integrated building automation system. Taking this approach would allow the company to reduce internal development costs—several existing applications will be replaced with the new system. The company could also achieve market expansion by entering new and emerging geographic markets and opening new sales channel in the form of VARs.
>
> *Key features*:
> - *Dynamic reconfiguration and field device management*: Support hardware devices from different manufacturers
> - *Alarm management*: Complies with regulations requiring covering life-critical systems, such as fire alarms, to operate within specific latency constraints
> - *User management, access control, and event management*: Integrate existing applications into a single unified software

6.6 Conclusions

This chapter explored the shaded portion of International Organization for Standardization/International Electrotechnical

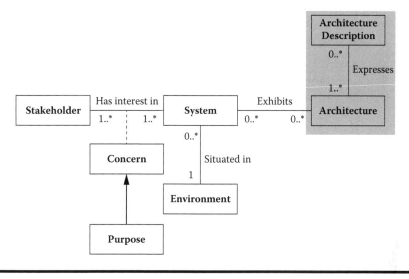

Figure 6.12   Conceptual framework for system and software architecture.

Commission (ISO/IEC) 42010 conceptual framework for system and software architecture shown in Figure 6.12. We demonstrated the use of the views-and-beyond approach for documenting the architecture and capturing its design rationale. Stakeholders have diverse concerns, and organizing the documentation as a set of views, in which each view addresses a specific set of concerns, helps them navigate the architecture documentation in a focused way to examine if their design concerns have been satisfied. The documentation example provided was intentionally kept brief, with emphasis placed on concepts learned in previous chapters. Those interested can explore the details of this approach in the work of Clements et al. (2010). Dickerson and Marvis (2009) discussed other architecture documentation frameworks (Department of Defense Architecture Framework [DoDAF] and Ministry of Defense Architecture Framework [MoDAF]) that are widely used in the defense community.

6.7 Questions

1. In a previous chapter, you created the architecture of the patient monitoring system. Create an architecture description document similar to the one discussed in this chapter for the patient monitoring system.
2. In a previous chapter, you modified the architecture of the building automation system to improve the availability of the virtual devices. Add another view to the architecture description document of the building automation system that captures your enhancements.

References

P. Clements, F. Bachmann, L. Bass, et al., *Documenting Software Architecture: Views and Beyond*, second edition. Upper Saddle River, NJ: Addison-Wesley, 2010.

C. Dickerson and D. Marvis, *Architecture and Principles of Systems Engineering*. Boca Raton, FL: CRC Press/Auerbach, 2009.

ISO/IEC/IEEE Systems and Software engineering—Architecture description, ISO/IEC/IEEE 42010:2011(E) (Revision of ISO/IEC 42010:2007 and IEEE Std 1471-2000), pp. 1–46, 2011.

Chapter 7

Architecture and Detailed Design

7.1 Introduction

The previous chapters explored the architectural analysis and design of the building automation system and showed that using the architecture-centric approach we arrive at an architecture that supports the business and mission goals of the application. This approach, however, leaves the fine-grained design details unspecified. Although such details are not a necessary part of architectural design, it is essential to understand the relationship between high-level architectural design and low-level detailed design and see how development work transitions from one to the other.

The mainstream methodology used in practice for detailed design is object-oriented analysis and design (OOAD) (Larman, 2005). Much of our discussion in Chapter 3 on achieving a broad functional understanding of a system forms an important part of OOAD as well. For instance, the natural starting point for OOAD is also a system context and is used for enumerating all of its actors. The goals of the actors in how they want to use the system become the basis for defining the use cases. The use cases are then elaborated to distill the functional responsibilities that a system must support to fulfill that use case. The elaboration is guided by a problem domain model that captures significant conceptual entities from the problem domain that are created, used, updated, or destroyed during the execution of the use cases. The problem domain model also becomes a basis for a standard vocabulary of significant concepts that are then used consistently for expressing use cases. OOAD, however, leverages this information for creating two additional artifacts: interfaces and a domain object model.

7.2 Defining Interfaces

We used the functional responsibilities derived in this manner to allocate responsibilities to individual components that were created as a part of elaborating the architecture in Chapter 5. For OOAD, the functional responsibilities derived from a use case are used for creating two types of classes for that use case: a *boundary class* that represents the interface that is used to interact with the components within the system and a request handler or a *controller class* that coordinates the invocation of components within the system to get the work done for the given use case. The actor initiates a use case through the boundary class (by generating an event, for instance), and the boundary class forwards the request from a user (the event that was generated, for instance) to the controller; the controller must then coordinate the invocation of a series of methods on components within the system in response to that event.

7.3 Creating the Domain Object Model

The problem domain model developed in Chapter 3 was used to create a standard vocabulary of terms that can be used in a consistent expression of all use cases. In OOAD, the domain model also becomes a motivator for creating classes known as *entity classes*. The entity classes represent significant concepts from the domain and house any important information associated with these concepts along with rules or logic for manipulating that information. They are created, used, updated, or destroyed during the execution of various use cases within the system and contain information of significant value. Therefore, entity classes together constitute a *domain object model* that is persisted in a persistent data store whose schema closely resembles this model.

The *domain object model* lives inside various components of the building automation architecture created in Chapter 5. For instance, it is natural for the *rule* domain object to reside in the *rule manager* component and for the *device* domain object to reside in the *virtual device* component. The components therefore may themselves play a role of a domain object manager or contain a *controller class* that manages access to these domain objects.

We capture the essence of this discussion in Figure 7.1. The figure shows that a user request is received through the boundary object that plays a role of an *interface,* which forwards the request to the controller object that plays a role of a *request handler,* which in turn invokes system operations on one or more objects that play a role of the *domain object manager,* which in turn invoke methods on one or more objects that play a role of *domain object*. The *request handler* eventually returns a system response to a user request.

We illustrate this approach for the use case scenario in which a field device generates a change-of-value event (such as someone raises the temperature property value of a thermostat on a building floor) that must be processed by a rule manager for taking an appropriate action (such as turning on the heating unit for that floor).

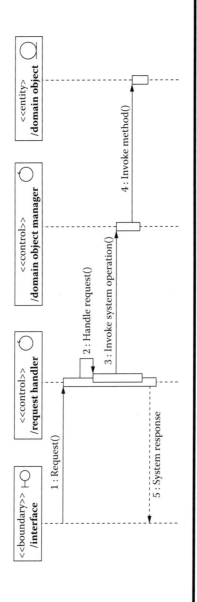

Figure 7.1 A conceptual pattern for detailed design.

7.3 The Rule Manager

Based on the conceptual pattern suggested in Figure 7.1, we consider the field device (i.e., the thermostat) to be the actor that generates a temperature change event that is first received by the virtual device component, a boundary class that plays the role of an interface into the building automation system. The virtual device publishes this event to the *publish-subscribe* connector, a controller class that plays the role of a request handler. It forwards the event to all those components that have subscribed to it. In our example, this happens to be the *rule manager*, a controller class that plays the role of a domain object manager. The rule manager invokes a rule evaluation method on one or more *rule* domain objects, and the results of this evaluation may suggest the rule manager further invoke a trigger method on one or more *command* domain objects to take the appropriate action. We show this conceptual sequence of actions in Figure 7.2.

This conceptual sequence of actions can serve as a good backdrop to our understanding, and we may be tempted to start putting together our blueprint for implementing the rule manager. There are, however, a few other considerations that need to be made.

The distinction between architecture design and detailed design is in the mind of the beholder. What may be detailed design to one may well be architecture design to another. So, when the detailed design of the rule manager is assigned to a project team, they may first go through the same architecture-centric process, namely, attribute-driven design (Bass et al., 2013), which was followed in the previous chapters for the building automation system, and create the internal architecture of the rule manager to address any of its architectural responsibilities before they address its functional responsibilities. Table 7.1 shows all of these responsibilities.

Table 7.1 is provided as an input to a project team, which they use to drive the design and development of the rule

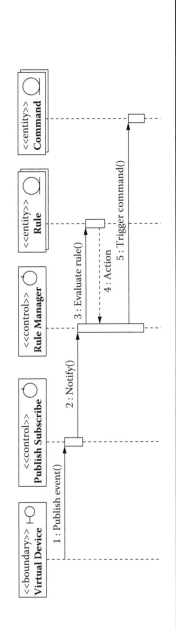

Figure 7.2 A conceptual sequence of actions triggered as a result of an event generated by a field device.

Table 7.1   Responsibilities of the Rule Manager

Responsibility	Type	Description
Process multiple concurrent events	Architectural	A building automation system can have configurations ranging from 500 to 500,000 field devices, which can generate considerable load on the system. To minimize the latency in processing events generated from these devices, they must be processed concurrently.
Evaluate and execute automation rules	Functional	Rules have a condition and an action. The condition specifies a threshold or value for one or more properties of one or more field devices that must be true for the action to be triggered. The action specifies commands that must be executed.
Send automation commands	Functional	The commands to be executed in response to an action must be dispatched to the appropriate field devices.

manager. We first create the internal architecture of the rule manager to address its architectural responsibilities. Then, we tackle the functional responsibilities.

7.3.1 Addressing Architectural Responsibilities

Our starting point would be the rule manager and its context (the other elements of the system with which it interacts) as presented in the final architecture of the building automation system. We show this in Figure 7.3.

For the rule manager to process multiple concurrent events, our primary performance design concern is how best to *manage the resources*. For this, we use a *schedule resources* tactic and decompose the rule manager into a subscription manager

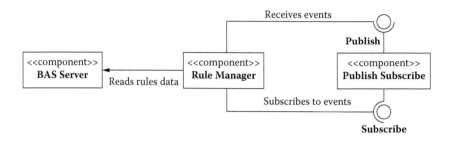

Figure 7.3   The rule manager and its context.

that subscribes to events received by the publish-subscribe connector and a priority event queue that receives these subscribed events and queues them in a priority order. We show the results of this decomposition in Figure 7.4.

Next, we use the *introduce concurrency* tactic to process multiple events simultaneously. We further decompose the rule manager and introduce a subcomponent *rule evaluator* that reads events from the event queue and selects appropriate *rules* to evaluate in response to the given events. If the conditions for these rules are met, then it triggers the action, which may involve executing one or more *commands*. The *command dispatcher* subcomponent handles the dispatching of these commands. We show the results of this decomposition in Figure 7.5.

The rule manager maintains a pool of rule evaluators and command dispatchers so they can process rules and commands concurrently. Some of these benefits, however, are mostly visible when multiple computational nodes are available.

Finally, we notice that the rule evaluator reads the rules from the Building Automation System (BAS) server. We use the *maintain multiple copies* tactic and create a rule cache to further improve the performance. The rule manager is responsible for maintaining this cache. Figure 7.6 shows the addition of a cache.

We address the functional responsibilities next.

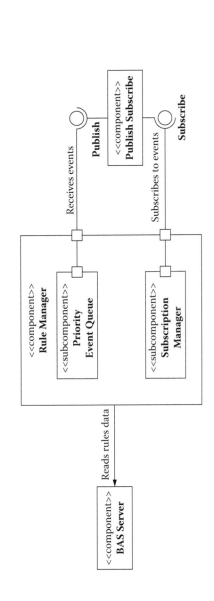

Figure 7.4 The rule manager after addressing the need to manage events to which it subscribes.

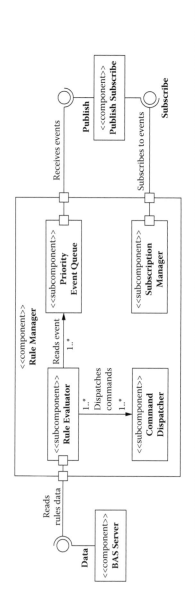

Figure 7.5 The rule manager after addressing the need to process concurrent events.

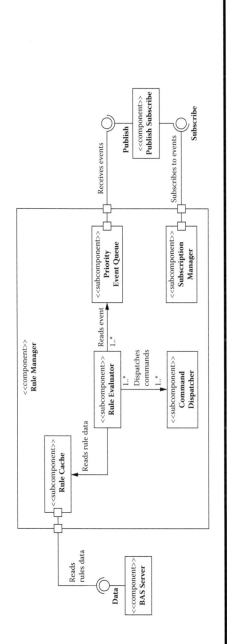

Figure 7.6   The rule manager after introducing a cache.

7.3.2 Addressing Functional Responsibilities

To handle evaluation and execution of automation rules, we need the capability to create and maintain these rules. To achieve this, we introduce a domain object rule and assign to it these responsibilities. Similarly, we introduce a domain object command that encapsulates all the information (such as the field device identifier, its properties and their values, etc.) required to execute a command. With these additions, Figure 7.7 shows the final architecture of the rule manager.

Figure 7.8 shows the interaction among the elements of the final architecture to handle an event generated by a field device. If we compare this figure with the conceptual diagram initially created in Figure 7.2, we see we have introduced several helper classes that address the performance design concerns related to the architecture responsibilities assigned to the rule manager.

7.4 Summary

The output at the end of architectural analysis and detailed design is a component architecture with its related design decisions traceable to the business goals and a substantial specification for system interfaces and domain and helper objects. System interfaces are a product of elaborating the use cases that describe the interaction of an actor at the boundary of a system and result in the creation of boundary and controller classes that manage this interaction. Domain objects are a product of the problem domain model that capture concepts from the problem domain of significant value. They result in creation of controller classes that manage the domain objects, entity classes that represent the domain objects, and the database schema used for persisting the domain objects. Helper classes result from addressing many of the architectural design concerns related to a component.

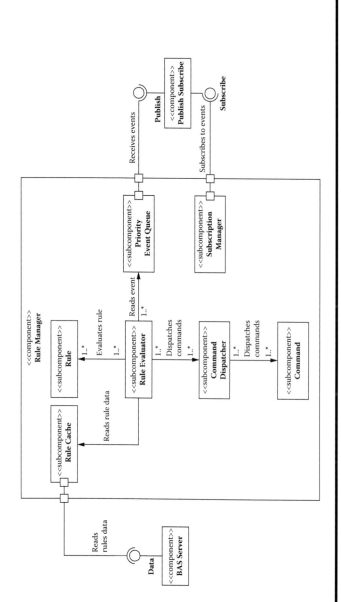

Figure 7.7 Final architecture of the rule manager after adding functional capabilities.

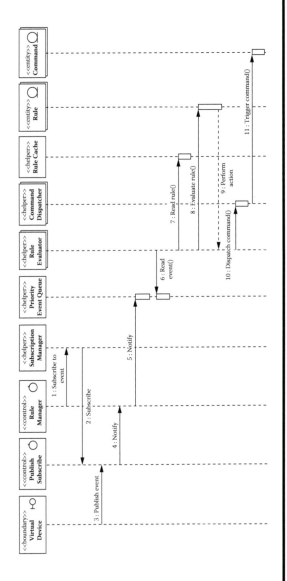

Figure 7.8   Handling of an event generated by a field device.

7.5 Question

1. Using the techniques described in this chapter, create a detailed design for
   a. Alarm manager
   b. Presentation manager
   c. Publish-subscribe connector

References

L. Bass, P. Clements, and R. Kazman, *Software Architecture in Practice,* third edition. Boston: Addison-Wesley, 2013.

C. Larman, *Applying UML and Patterns*, third edition. Upper Saddle River, NJ: Prentice-Hall, 2005.

Chapter 8

Role of Architecture in Managing Structural Complexity

8.1 Introduction

In the previous chapters, we learned how to drive the development of an architecture of a system. Embracing architecture-centric practices results in a system that not only meets its quality expectations but also can lead to a significant reduction in excessive complexity, suggesting that excessive complexity may be an incidental by-product of development methodologies that lack focus on the systemic properties of a system that have a strong influence on its architecture. Complexity in a system is related to the degrees of cohesion and coupling within and between a system's modules (a system's cohesive units of implementation). Systems designed with high cohesion achieve better separation of concerns with each module within the system addressing one and only one concern. High cohesion also leads to low coupling, by which interdependencies among the modules are significantly reduced.

8.2 Analyzing System Complexity

Complexity is perceived as the number of interacting elements within a system, the internal structure of these elements, and the number and nature of interdependencies among them (Taylor et al., 2009). At the design level, complexity is related to the degrees of cohesion and coupling within and between a system's elements (Stevens et al., 1974). Design-level cohesion and coupling in a system can be analyzed using a design structure matrix or DSM (Eppinger, 1991; Steward, 1981). A DSM is also known as a dependency structure matrix, dependency source matrix, and a dependency structure method.

DSM is a square matrix (it has an equal number of rows and columns) that maps dependencies or relationships among elements of a system. All elements appear in both the rows and the columns, and dependencies are signaled at the intersection points of the items in the matrix. For example, Figure 8.1a shows a DSM for a software system that has been decomposed into four modules: A, B, C, and D. The rows and columns of the matrix represent the same modules. The dependencies of a module are read down a column. For instance, reading down the first column, we can see that module A depends on module C. We also see that module A does not depend on modules B or D as those cells are empty. These dependencies are marked in Figure 8.1b and are known as *uses* dependencies (as in module A *uses* module C). Reading across a row gives us dependencies on a module. For instance, reading across the first row, we can see that modules C and D depend on module A. These dependencies are marked in Figure 8.1c and are known as *used by* dependencies (as in module A is *used by* modules C and D). The identity diagonal represents a dependency of a module on itself.

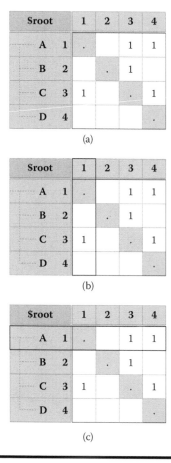

Figure 8.1 A DSM showing (a) dependencies among software modules, (b) the uses dependencies, and (c) the used by dependencies.

A DSM provides a simple, compact, and visual representation of a system. DSMs to date have been used not only in component-based and architecture decomposition and integration analysis but also in organization, project, product planning, and management contexts (Lindemann, 2009). The use of DSMs in software engineering has mostly focused on understanding design rules (Sullivan et al., 2001) and has been increasingly incorporated into reverse engineering tools and tools to develop architectures such as Lattix (2011) and Structure 101 (2011).

$root	1	2	3	4
Module A 1	.		6	9
Module B 2		.	19	
Module C 3	7		.	8
Module D 4				.

Figure 8.2   DSM showing dependency strength between elements of a system.

The DSM in Figure 8.1 is a binary matrix; its cells are populated with zeros and ones. Binary matrices are useful because they show presence or absence of a relationship between pairs of elements of a system. As shown in Figure 8.2, a matrix can also show weighted dependencies. For instance, reading down column 1, we can see that module A depends on module C with a dependency strength of 8.

The presence of a large number of dependencies within a cell may be indicative of poor cohesion. For instance, reading down column 3 and across row 2, we find module C has 19 dependencies on module B. Although 19 is not a large number, imagine if it were in the hundreds or more. The presence of such a large number would suggest that B was performing many (perhaps diverse and unrelated) tasks for which C must rely on it.

8.2.1 *Partitioning a DSM*

We can reorder the subsystems in the DSM through a process called *partitioning*. The objective of partitioning is to sequence the DSM rows and columns such that the new DSM arrangement does not contain any feedback loops or cycles, thus transforming the DSM into a *lower triangular* form. A feedback loop or cycle exists when there are circular dependencies (for instance, an element X depends on an element Y and element Y in turn depends on element X). For complex systems, it is highly unlikely that simple row and column

$root		1	2	3	4
Module D	1	.			
Module A	2	9	.	6	
Module C	3	8	7	.	
Module B	4			19	.

Figure 8.3   Reordered DSM.

manipulation will result in a lower triangular form. Therefore, the objective changes from eliminating the feedback loops to moving them as close as possible to the diagonal (this form of the matrix is known as *block triangular*).

Figure 8.3 shows this reordering for the DSM in Figure 8.2. The reordering yields a block triangular matrix consisting of three groups or blocks, the first consisting of module D, the second consisting of modules A and C (because they are coupled together due to circular dependency), and the third consisting of module B. Note that there are no dependencies (and hence no cycles) in the area of the matrix above the marked blocks.

It is equally possible to learn about which elements of the system might possibly have to be reworked (e.g., split into two elements or perhaps merged) to achieve better architecture. For example, Figure 8.4 shows combining of modules A and C into a compound module A-C (note that a module is simply an abstraction, a highly cohesive unit of implementation; in

	$root		1	2	3	4
	Module D	1	.			
A-C	Module A	2	9	.	6	
	Module C	3	8	7	.	
	Module B	4			19	.

Figure 8.4   Merging modules.

$root		1	2	3
Module D	1	.		
Module A-C	2	17	.	
Module C	3		19	.

Figure 8.5 Collapsed hierarchy.

a software system, this can be as small as a single class or as large as an entire subsystem containing thousands of classes).

As Figure 8.5 shows, we can now collapse module A-C so that we do not see modules A and C. The dependency that module D has on module A-C is an aggregation of the dependencies that module D had on module A and module C. By default, the aggregation is a simple summation, although other approaches to aggregation can also be configured. Notice the resulting matrix is now a lower triangular matrix. There are no dependencies (and hence no cycles) in the upper area of the matrix above the diagonal.

Partitioning can be done manually or by invoking a partitioning algorithm. The overall goal of DSM partitioning algorithms is to order a system starting with modules that use most of the other modules and ending with those that provide most to the others while grouping modules that are highly interdependent together in the ordering. DSMs thus can give us a sense of which modules underlie other modules and how coupled and cohesive they are. For example, the system represented by the DSM in Figure 8.5 can be envisioned as a layered system (shown in Figure 8.6) with the first block as the top layer, followed by the second block, and finally the third block as the bottom layer.

8.2.2 Partitioning Algorithms

There are several DSM partitioning algorithms. They are mostly similar except in how they identify cyclic

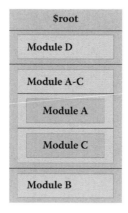

Figure 8.6   Layered view of the system.

dependencies. All partitioning algorithms proceed as follows (Lindemann, 2009):

1. Identify system elements that have no dependencies from the rest of the elements in the matrix. These elements can easily be identified by observing an empty row in the DSM. Place those elements to the left and top of the DSM. Remove them from further consideration.
2. Identify system elements that have no dependencies on other elements in the matrix. These elements can easily be identified by observing an empty column in the DSM. Place those elements to the right and bottom of the DSM. Remove them from further consideration.
3. If after steps 1 and 2 there are no remaining elements in the DSM, then the matrix is completely partitioned; otherwise, the remaining elements contain cycles (at least one).
   a. There are many ways of determining a cycle, but one such method is called *path searching*. In path searching, dependencies are traced either backward or forward from a given element until the element is encountered twice. All elements between the first and second occurrence of this element constitute a cycle.

b. Group the elements involved in a single cycle into one representative element and go to step 1.

Once elements are grouped, a topological sort can be used to order them so that dependencies between the groupings go in the same direction.

EXAMPLE 8.1

Figure 8.7 shows how partitioning works:

- Figure 8.7a shows the original dependency matrix.
- Figure 8.7b shifts element F to the left and top because it has no dependencies (empty row); we ignore F's row and column in subsequent analysis.
- Figure 8.7c shifts element E to the left and top because it has no dependencies (empty row); we ignore E's row and column in subsequent analysis.
- Figure 8.7d shifts element A to the right and bottom because it does not depend on anything (empty column); we ignore A's row and column in subsequent analysis.
- Figure 8.7e shifts element B to the right and bottom because it does not depend on anything (empty column); we ignore B's row and column in subsequent analysis.
- The remaining elements (C and D) have a cycle, so we collapse them into a single representative element called C-D. There are no more elements remaining at this point, and Figure 8.7f shows a completely partitioned matrix. Notice that the matrix is in a lower triangular form at this point.

8.2.3 Tearing a DSM

Tearing is the process of choosing the set of dependencies that create feedback loops that, if removed from the matrix (and then the matrix is repartitioned), will render the matrix lower triangular. The dependencies that we remove from the matrix are called "tears." No optimal method exists for tearing, but Lindemann

$root		1	2	3	4	5	6
A	1	.			1		
B	2		.	1			
C	3			.	1	1	1
D	4			1	.	1	
E	5					.	1
F	6	1					.

(a)

$root		1	2	3	4	5	6
F	1	.					
A	2		.			1	
B	3			.	1		
C	4	1			.	1	1
D	5				1	.	1
E	6	1					.

(b)

$root		1	2	3	4	5	6
F	1	.					
E	2	1	.				
A	3			.			1
B	4				.	1	
C	5	1	1			.	1
D	6		1			1	.

(c)

Figure 8.7 (a) Original DSM; (b) F shifted left and top; (c) E shifted left and top; *continued*

$root		1	2	3	4	5	6
F	1	.					
E	2	1	.				
B	3			.	1		
C	4	1	1		.	1	
D	5		1		1	.	
A	6					1	.

(d)

$root		1	2	3	4	5	6
F	1	.					
E	2	1	.				
C	3	1	1	.	1		
D	4		1	1	.		
B	5			1		.	
A	6				1		.

(e)

$root		1	2	3	4	5
F	1	.				
E	2	1	.			
C-D	3	1	2	.		
B	4			1	.	
A	5			1		.

(f)

Figure 8.7 (continued) (d) A shifted right and bottom; (e) B shifted right and bottom; (f) Partitioned DSM.

(2009) recommend keeping the tears to a minimum (also sometimes referred to as a *minimum feedback set*).

8.3 Managing Structural Complexity

We have learned that quality attribute requirements have a significant impact on the architecture of a system. If not given adequate attention early in the development life cycle, it becomes extremely difficult to achieve these requirements without significantly altering the structure of the system. When faced with such a situation, ad hoc approaches are typically used to retrofit the existing structure for achieving these qualities rather than significantly altering it in any way. For example, when faced with a performance requirement in a software system, many developers may use layer bridging within a perfectly layered system. Over time, such approaches introduce arbitrary complexity into a system, making it increasingly complicated and ultimately requiring a major restructuring during the course of its evolution.

Without a disciplined design approach, such as the one prescribed in this text, design concerns are likely to be comingled and not addressed systematically. In the worst case, a huge monolithic element, within which all issues are addressed, is naturally formed. Fixing problems under such situations requires considerable effort. In software systems, it is not unusual for maintainers to insert their pieces of code into different blocks of the same program. It is also not uncommon that a lot of obsolete code remains because people sometimes forget which parts belong to them and are reluctant to delete others' work. Accordingly, the system continues to become bloated, making it difficult to maintain and evolve.

When better designed, concerns can be separate, and the system can be more modular. We still, however, need to carefully manage the associations among the modules of a system; otherwise, they can become coupled inadvertently. When there is no rule to guide the relationship between elements, the ripple effect of change in one element may extend to a significant part of the system.

> With a well-designed architecture, there is a scaffold to guide the development of a system. Elements are restricted within fixed boundaries and are provided with few interfaces to allow communicating with outsiders.

8.3.1 Testing the Hypothesis

To test the hypothesis that an architecture-centric approach can lead to better separation of concerns and therefore lower structural complexity, a simple experiment was conducted. The search was for a project that had been abandoned, after several successful releases, because its developers were finding it increasingly hard to extend and maintain the application once it was operational (Sangwan et al., 2008; Sangwan and Neill, 2009). It was an Internet-based collaborative work environment incorporating speech communications, videoconferencing, e-mail, file sharing, and a shared whiteboard that had experienced the typical scope creep of operational systems as each new feature had been added successfully by the development team. As the system evolved, however, performance and reliability problems emerged. The system experienced significant degradation of performance when many shared whiteboards were active and would freeze if network connectivity were compromised. Despite rewriting thousands of lines of code, the team failed to overcome these issues and eventually abandoned the project.

Analysis of the abandoned system revealed that systemic concerns such as performance and reliability were distributed throughout the application to such an extent that improving them was now an impossible task. This was not merely a case of poor initial design, however. On the contrary, the initial architecture exhibited good object-oriented design. Given the requirements of the system, we would anticipate separate components for media, mail, drawing, and so on, and that is what we found.

So, what went wrong? As is typical in object-oriented analysis, the primary driver of design is the use case model. This promotes functional and data requirements, and this is reflected in the resulting architecture. Nonfunctional requirements are treated as secondary concerns that all components should exhibit, but when those requirements are not met, there is no single source of the problem to be repaired—the flaw is distributed across the application. In a general sense, although functional requirements can be achieved in any structure, only certain structures enable the qualities desired from those functions or the system as a whole.

The failure of the collaborative system project can be attributed to treating its significant nonfunctional requirements, such as performance and reliability, as secondary, such that it became problematic to address these concerns late in the development life cycle.

In the case of performance, for instance, event handling was recognized as one of the factors with an impact on processing efficiency; drawing, erasing, and mouse dragging generated thousands of events that most of the time led to unnecessary updates to the whiteboards. Keeping a history of these events so that parts of the session could be replayed led to inefficient memory utilization. In the case of reliability, breaks in Internet connectivity during periods of inactivity were an issue. These breaks were not discovered until buffers of the whiteboards began to overflow because of unsent data, further cascading into other failures.

If the issues were recognized, what made it problematic to fix them? The code for performance and reliability concerns was spread throughout the system. Separating these concerns and retrofitting the existing structure to meet the performance and reliability demands led to the introduction of arbitrary complexity (undesirable dependencies, cycles, paths of communication, fat methods and classes, etc.), making the system excessively complex or complicated. This made it difficult to understand, maintain, test, and troubleshoot.

Figure 8.8 shows the DSM for the failed system. As can be seen from this DSM, there are two significant clusters marked as tangle of 9 and tangle of 3 in the figure. There are numerous cells in the DSM for which the number of dependencies is in the hundreds or thousands. The clusters are indicative of extensive coupling, and large numbers of dependencies are indicative of poor cohesion. Together, they suggest a system that not only would be difficult to understand but also would be difficult

		order	member	client	server	ui	store	data	draw	skin	mail	voip	group	media	net	file	util
	order			4	5												
	member	4		3													
	client	15	13					14									
	server	14															
	ui	30	355	2184			2			3	2						
	store			59	1	7		13									
	data	41	225	275	237	21	3										
	draw			323	35												
	skin			84													
	mail			95	44												
	voip			211													
	group		59	612	195	159	39	5				7					
	media			169										22			
	net	194	199	1513	2146	36	127	158			116		233	11		70	32
	file			140	108		4	6			46		14		1		1
	util	74	70	1490	622	133	31	75	26	34	144	102	88	84	159	96	

Figure 8.8 DSM for the failed Internet-based collaboration system. (Adapted from R. Sangwan and C. Neill, Characterizing essential and incidental complexity in software architectures. In *Proceedings of the 8th Working IEEE/IFIP Conference on Software Architecture*, August 14–17, 2009, Cambridge, UK.)

to maintain as any change made to the system had the likelihood of propagating to a significant proportion of the system. Detailed examination of the system revealed that the coupled and incohesive part made up 39% of the entire code base of the system (348 of the 897KLOC [Thousand Lines of Code]).

To examine if its complexity was a by-product of the development method, we chose to develop a new version of the system employing the architecture-centric approach advocated in the text. Rather than focus on the entire system, we chose the whiteboard subsystem of the Internet-based collaboration environment because this was the part that suffered most from poor modularity, performance degradation, and lack of reliability—creating conditions that were hard to troubleshoot while the system was in operation. The whiteboard subsystem is represented by the tangle of 9 (in the upper left) of the DSM shown in Figure 8.8.

Table 8.1 shows a set of scenarios we chose to drive the design. Starting with a monolithic subsystem, we addressed driver 1 and applied the modifiability tactics to limit the impact of change and minimize the number of dependencies on the parts of the subsystem responsible for integrating new input or display devices. Figure 8.9 shows we first introduced a Whiteboard Module and decomposed it into a BoardSensor and a BoardPane to handle disparate input and output devices, respectively, a tactic to *anticipate expected changes*. We added a BoardConnection element to handle communication between various input and output devices, a *restrict communication paths* tactic. Finally, a BoardInterface element was introduced for connecting different BoardPane elements to different BoardSensor elements, an *intermediary* tactic.

Figure 8.9 uses the layered view convention first introduced in Figure 8.6. The same convention is also used for subsequent figures related to this design experiment.

Next, driver 2 was addressed using performance tactics (*control sampling frequency* and *reduce computational overhead*) to add support for critical system operations so

Table 8.1   Prioritized Architectural Drivers

Driver No.	Quality Attribute	Architectural Drivers
1	Modifiability	The system shall be capable of receiving inputs from different devices, such as a wireless pen tablet or touch screen, and sending output to different devices, such as a display station or a smartphone.
2	Performance	The system shall respond within no more than 1 s for sketching or drawing and no more than 5 s for any other operation.
3	Testability	Tracking progress of execution and troubleshooting any failures shall be possible at runtime.
4	Availability	The system shall recover from unstable network connections.

Source: Adapted from R. Sangwan, L. Lin, and C. Neill, *IEEE Computer* October 2008, pp. 99–102.

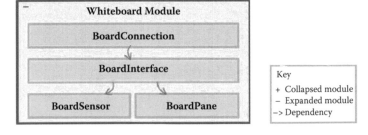

Figure 8.9   Whiteboard subsystem after adding support for disparate input and output devices. (Adapted from R. Sangwan, L. Lin, and C. Neill, *IEEE Computer* October 2008, pp. 99–102.)

they operate within specific latency constraints. The Event Management Module was created to manage sampling of events (from drawing, erasing, mouse dragging, etc.) and implementing efficient algorithms for updating the whiteboard display (BoardPane) when events arrive. This module was also

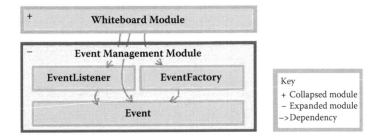

Figure 8.10   Whiteboard subsystem after adding support for operations to perform within specified latency constraints. (Adapted from R. Sangwan, L. Lin, and C. Neill, *IEEE Computer* October 2008, pp. 99–102.)

responsible for implementing a memory-efficient history mechanism for these events so sessions could be replayed. The resulting system is shown in Figure 8.10. The *EventFactory* is responsible for creating the different types of *Events* that the *WhiteBoard Module* generates, and the *EventListener* listens for, and controls sampling of, these events.

The testability tactic (*built-in monitor*) is applied next to address driver 3 to minimize the effort required to test the system at runtime. A Logging Module is added with a Logger element used for publishing information and Reader element used for providing the published information to its subscribers. A Testing Module is added with a Server element, which can output logged information over the Internet to any telnet console compatible with the Transmission Control Protocol/Internet Protocol (TCP/IP), and a Console element that can display the logged information. The results are shown in Figure 8.11.

Finally, reliability (driver 4) is achieved by creating a Messaging Module with the Messenger element used for reading and writing messages over the Internet and the Message element used for storing and manipulating these messages. The Messenger element implements a *heartbeat* tactic for fault detection (e.g., broken network connection) and notifies the BoardPane element of the Whiteboard Module. The BoardPane element recovers from the fault by collaborating

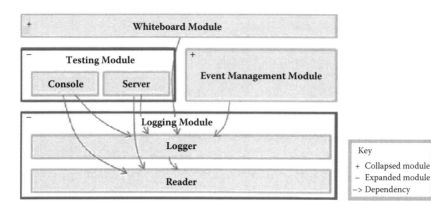

Figure 8.11   Whiteboard subsystem after adding support for minimizing the effort to test the system while it is in operation. (Adapted from R. Sangwan, L. Lin, and C. Neill, *IEEE Computer* October 2008, pp. 99–102.)

with the *Messenger* element using a *state resynchronization* tactic for fault recovery (e.g., resynchronize content after a network connection is reestablished).

As can be seen in the final subsystem architecture, all dependencies between layers flow in the same direction, from the topmost layer downward. This indicates that there are no cycles in the dependency graph and therefore no tangles that would propagate changes and errors throughout the system. This is shown in the DSM in Figure 8.12.

8.4 Discussion and Conclusions

Quality attributes are critical to the evolution of a software architecture. If put into the design earlier, they make the resulting system less complex and more stable, like the newly devised whiteboard subsystem presented in this work. Not paying attention to these qualities can make the system so tangled that it becomes difficult to manage, like the original whiteboard subsystem.

	Whiteboard	Event mgmt	Testing	Messaging	Logging
⊞ Whiteboard					
⊞ Event mgmt	88				
⊞ Testing	1				
⊞ Messaging	15	39	28		
⊞ Logging	64	30	30	27	

Figure 8.12   DSM for the reengineered whiteboard subsystem. (Adapted from R. Sangwan and C. Neill, Characterizing essential and incidental complexity in software architectures. In *Proceedings of the 8th Working IEEE/IFIP Conference on Software Architecture*, August 14–17, 2009, Cambridge, UK.)

The case study work stepped through the significant quality attributes for the whiteboard subsystem by adopting appropriate patterns and tactics and derived the new architecture for the same system with clear-cut internal structure. Modifiability, performance, testability, and reliability were achieved while effectively managing structural complexity.

8.5 Discussion Questions

1. Given the following dependency structure matrix for a system consisting of modules A, B, C, D, and E,

	A	B	C	D	E
A	.	X	X	X	X
B		.	X	X	X
C		X	.	X	X
D				.	X
E					.

    a. Create a block triangular matrix.

    b. From the block triangular matrix, create a layered view of the system.

2. Given the final architecture of the building automation system in Figure 5.7,

    a. Create a DSM.

    b. Transform the DSM into a block triangular form.

    c. Create a layered view of the system using the block triangular form of the matrix.

References

S. Eppinger, Model-based approaches to managing concurrent engineering, *J. Eng. Design* 2 (4), 283–290, 1991.

Lattix, version 7.0. 2011. http://www.lattix.com.

U. Lindemann, *Technical DSM Tutorial*. 2009. http://dsmweb.org.

R. Sangwan, L. Lin, and C. Neill, Structural complexity in architecture-centric software evolution, *IEEE Computer* October 2008, pp. 99–102.

R. Sangwan and C. Neill, Characterizing essential and incidental complexity in software architectures. In *Proceedings of the 8th Working IEEE/IFIP Conference on Software Architecture*, August 14–17, 2009, Cambridge, UK.

W. Stevens, G. Myers, and L. Constantine, Structured design. *IBM Syst. J.* 13 (2), 115–139, 1974.

D.V. Steward, The design structure system: A method for managing the design of complex systems, *IEEE Trans. Eng. Manage.* 28 (3), 71–74, 1981.

Structure 101, Version 3.5, Build 1527. 2011. http://www.headway-software.com.

K.J. Sullivan, W.G. Griswold, Y. Cai, and B. Hallen, The structure and value of modularity in software design, *ACM SIGSOFT Softw. Eng. Notes* 26 (5), 99–108, 2001.

R. Taylor, N. Medvidovic, and E. Dashofy, *Software Architecture: Foundations, Theory and Practice*. Hoboken, NJ: Wiley, 2009.

Index

Printed and bound by CPI Group (UK) Ltd, Croydon, CR0 4YY

23/10/2024

01777673-0004